Springer Series in Optical Sciences

Volume 172

For further volumes:
http://www.springer.com/series/624

Springer Series in Optical Sciences

The Springer Series in Optical Sciences, under the leadership of Editor-in-Chief William T. Rhodes, Georgia Institute of Technology, USA, provides an expanding selection of research monographs in all major areas of optics: lasers and quantum optics, ultrafast phenomena, optical spectroscopy techniques, optoelectronics, quantum information, information optics, applied laser technology, industrial applications, and other topics of contemporary interest.
With this broad coverage of topics, the series is of use to all research scientists and engineers who need up-to-date reference books.
The editors encourage prospective authors to correspond with them in advance of submitting a manuscript. Submission of manuscripts should be made to the Editor-in-Chief or one of the Editors. See also www.springer.com/series/624

Mohamed Atef · Horst Zimmermann

Optical Communication over Plastic Optical Fibers

Integrated Optical Receiver Technology

Springer

Mohamed Atef
Institute of Electrodynamics,
Microwave and Circuit Engineering
Vienna University of Technology
Vienna
Austria

and

Faculty of Engineering
Department of Electrical Engineering
Assiut University
Assiut
Egypt

Horst Zimmermann
Institute of Electrodynamics,
Microwave and Circuit Engineering
Vienna University of Technology
Vienna
Austria

ISSN 0342-4111 ISSN 1556-1534 (electronic)
ISBN 978-3-642-30387-6 ISBN 978-3-642-30388-3 (eBook)
DOI 10.1007/978-3-642-30388-3
Springer Heidelberg New York Dordrecht London

Library of Congress Control Number: 2012947394

Printed on acid-free paper

Springer is part of Springer Science+Business Media (www.springer.com)

Preface

The Plastic Optical Fiber (POF) provides benefits compared to glass fiber. The best known POF is the step-index (SI) PMMA POF with a core diameter of 1 mm. It is simpler and less expensive than glass fibers. Large-core POF has greater flexibility and resilience to bending, shock, vibration, and it is easier in handling and connecting. The transmission windows of the POF are in the visible range. These advantages make POF very attractive for use in short-range communication and within in-building networks.

The cheap standard PMMA SI-POF has the lowest bandwidth and the highest attenuation among multimode fibers. This small bandwidth limits the maximum data rate which can be transmitted through POF. To overcome the problem of its high transmission loss, very sensitive receivers must be used to increase the transmitted length over POF. The POF's limited bandwidth problem can be decreased by using multilevel signaling like Multilevel Pulse Amplitude Modulation (M-PAM) or Discrete Multi Tone (DMT). Another method is to use equalization techniques to reduce the effect of limited bandwidth.

This book presents integrated optical receivers fabricated with a low-cost Si process and having a good performance with multilevel modulation. These receivers achieve a good transmission performance over standard PMMA SI-POF (higher sensitivity and higher data rate transmission over longer POF length). Also, fully integrated optical receivers with an integrated POF equalizer achieving free of error 1.25 Gbit/s transmission over 50 m SI-POF will be introduced.

In the design of the multilevel signaling optical receiver a linear transimpedance amplifier with an automatic gain control and no limiting amplifiers are used to have equally spaced output signal voltage levels. For many applications it is desirable to integrate the photodiode with a transimpedance amplifier and the post amplifiers into the same chip. Placing the transimpedance amplifier adjacent to the photodiode improves the performance by reducing lead capacitance and sensitivity to interference, thereby giving higher speed and lower noise. A further advantage is the reduction of the external circuitry required. Hence, overall cost and PCB board size can be reduced.

Experiments regarding the application of M-PAM over the SI-POF channel are presented using the integrated optical receivers introduced. The presented receivers are able to receive a free of error signal at a data rate of 1 Gbit/s over 20 m SI-POF, 400 Mb/s over 50 m SI-POF, and 170 Mbit/s over 115 m of SI-POF. All of the reached results are free of errors ($<10^{-8}$), were obtained in real time and have a data rate-length product of 20 Gb/s m.

A single-chip optical receiver with an integrated equalizer is used to achieve a high performance gigabit transmission over step-index plastic optical fiber (SI-POF). The integrated equalizer can compensate for different POF lengths up to 50 m. The integrated optical receiver is fabricated in a low-cost silicon 0.6 µm BiCMOS technology and has a power consumption of 100 mW. Real-time transmission at data rates of 1.8 Gbit/s over 20 m SI-POF and 1.25 Gbit/s over 50 m SI-POF with high sensitivities and BER of 10^{-9} is achieved.

This book has eight chapters and is organized as follows:

In the beginning of this book in Chap. 1, the motivation for optical communication over step index plastic optical fibers and the use of fully integrated optical receiver technology are discussed.

Chapter 2 provides the description of different multilevel signaling modulation formats and gives a brief comparison between multilevel and binary signaling. We motivate the importance of the multilevel signaling when the system frequency response has a certain roll-off. Some issues of the different methods to generate and decode the M-PAM are introduced. The calculation of symbol error rate for M-PAM and BER for binary signal will be provided.

In Chap. 3 different types of equalizations are described. Preequalization, post-equalization, electronics and analogue equalization are compared. The implementation of different kinds of equalizers is also discussed in this chapter. Performance comparison between different types of equalization techniques for high-speed communication over SI-POF will be addressed.

The first part of Chap. 4 explains the importance and benefits of PMMA SI-POF compared to copper wire and glass fiber. The main applications of PMMA SI-POF will be introduced. The characteristics of PMMA SI-POF like frequency response, bending loss and coupling loss will be discussed. The next sections in Chap. 4 will be about the available light sources for PMMA POF and how to minimize the effects of POF modal dispersion. The last section contains the state of the art for high speed transmission systems over standard SI-POF.

Chapter 5 provides the theoretical background of silicon PIN photodiodes. The design of large-area integrated PIN photodiode in silicon BiCMOS technology will be described. A brief comparison between this photodiode design example and other photodiode designs will be introduced.

Chapter 6 discusses the theory of transimpedance amplifiers (TIAs). A detailed analysis for the frequency response and noise of transimpedance amplifiers are shown. Different TIA topologies and their performance will be discussed. The end of the chapter contains a discussion about two types of post amplifiers, limiting amplifiers and automatic gain control amplifiers.

The first part of Chap. 7 discusses the circuit structure of the multilevel signaling integrated optical receiver and its importance. The next sections give circuit descriptions and experimental results for the fabricated integrated optical receivers with different bandwidth and different sensitivities. The experimental results for binary, 4-PAM and 8-PAM signal transmission over PMMA SI-POF lengths from 10 m to 115 m are also included.

Chapter 8 gives circuit descriptions and experimental results for the integrated adjustable POF equalizer. Fully integrated optical receivers with integrated equalizer will be introduced. The experimental results for equalized signals over different SI-POF lengths are also included for data rates up to 1.25 Gbit/s. A comparison to recently published results of integrated POF equalizers will be presented at the end of the chapter.

Chapter 9 compares receivers and transmission results of multilevel signaling and of binary modulation. This chapter finally draws some conclusions.

Vienna, 2012

Mohamed Atef
Horst Zimmermann

Acknowledgments

The authors want to express sincere appreciations to all of their colleagues in the circuit engineering group who offered the opportunity for many interesting discussions. Special thanks are directed to Robert Swoboda, who made several of his receiver designs available, and to Wolfgang Gaberl, who shared his valuable knowledge and helped in the measurements as well as with the simulation tools.

We extend our sincere thanks to Dr. Ascheron and his team at Springer for their good cooperation and their technical support with the text processor.

Vienna, 2012

Mohamed Atef
Horst Zimmermann

The scholarship from Egyptian Ministry of Higher Education, which made my stay at Vienna University of Technology and the research on integrated optical receivers and POF data transmission possible, is gratefully acknowledged.

Furthermore, I am grateful to my wife "Asmaa Hosni" for her support and patience during my work.

Vienna, 2012

Mohamed Atef

Contents

1 **Introduction** 1

2 **Multilevel Signaling** 5
2.1 Introduction 5
2.2 Multilevel Signaling and Channel Capacity 6
2.3 Optimum Number of PAM Levels 10
2.4 M-PAM Symbol Error Rate 11
2.5 Bit Error Rate for Binary Signals 13
2.6 M-PAM Generation and Decoding 14
2.6.1 M-PAM Generation 14
2.6.2 M-PAM Decoding 19
2.7 DC Balance Code 19
2.8 Multilevel Signaling Related Work 20

3 **Equalization Techniques** 23
3.1 Passive Equalizer 24
3.2 Active Equalizer 24
3.2.1 Source Degeneration 24
3.2.2 Differential Source Degeneration Adjustable Equalizer 26
3.2.3 Inductive Load Amplifier 26
3.3 Filter Implementation Using Transversal and Lattice Structures 27
3.4 Linear Equalization 28
3.4.1 Continuous Time FIR Filter Implementation 28
3.4.2 Discrete Time FIR Filter Implementation 30
3.5 Equalization Algorithms 32
3.6 Adaptive Equalization 34
3.6.1 Continuous Time Adaptive Equalizer 34
3.6.2 Discrete Time Adaptive Equalizer 34

3.7 Nonlinear Equalization 36
3.7.1 Decision Feedback Equalizers (DFE) 36
3.7.2 MLSE Nonlinear Equalization 37
3.8 Performance Comparison of Equalization Methods 39

4 High-Speed Transmission over Step-Index PMMA Plastic Optical Fibers 41
4.1 Why Large-Core PMMA POF? 41
4.2 POF Main Applications 42
4.3 Characteristics of Step-Index PMMA POF 43
4.3.1 POF Frequency Response and Bandwidth 43
4.3.2 POF Bending Loss 45
4.3.3 POF Coupling Loss 46
4.4 Light Sources for PMMA POF 47
4.4.1 Light Source Modulation 48
4.4.2 Laser Diode Modulation 49
4.4.3 Laser Nonlinearity 50
4.5 Minimizing the Effects of POF Modal Dispersion 50
4.6 State of the Art for Transmission over Standard PMMA SI-POF 52
4.6.1 Without Using Equalization 52
4.6.2 By Using Equalization 55

5 Integrated Photodiode 61
5.1 Optical Absorption 61
5.2 Photodiode Responsivity 63
5.3 Quantum Efficiency 63
5.4 Carrier Drift and Diffusion 64
5.5 PIN Photodiode Capacitance 65
5.6 Photodiode Speed 65
5.7 Structure of the Implemented Photodiode 66

6 Transimpedance Amplifier 69
6.1 Introduction 69
6.2 Transimpedance Amplifier 69
6.3 Shunt-Shunt Feedback TIA 70
6.3.1 Frequency Response 71
6.3.2 TIA Noise Analysis 74
6.3.3 TIA with Common-Emitter Input Stage 75
6.3.4 TIA with Common-Source Input Stage 77
6.3.5 Multistage Inverter Based CMOS TIA 78
6.4 Common-Base/Gate Input Stage 79
6.5 Regulated-Cascode TIA 80
6.6 Inverter Based Common-Drain Feedback TIA 81

6.7 TIA with Gain Control . . . 83
6.8 Post Amplifier . . . 84
6.8.1 Limiting Amplifier . . . 84
6.8.2 AGC Amplifier . . . 85
6.9 Dynamic Range . . . 86

7 Integrated Optical Receiver and Multilevel Transmission over PMMA SI-POF . . . 87
7.1 Optical Receiver with 622 MHz Bandwidth . . . 87
7.1.1 Circuit Description . . . 87
7.1.2 Experimental Results . . . 89
7.1.3 Gigabit Transmission over SI-POF . . . 95
7.1.4 1 Gbit/s over 20 m POF . . . 96
7.2 Optical Receiver with 200 MHz Bandwidth . . . 97
7.2.1 Experimental Results . . . 97
7.2.2 500 Mbit/s Transmission over SI-POF . . . 100
7.3 High Linearity Optical Receiver with 112 MHz Bandwidth . . . 101
7.3.1 Circuit Description . . . 103
7.3.2 Experimental Results . . . 105
7.3.3 500 Mbit/s Transmission over 40 m SI-POF . . . 110
7.3.4 400 Mbit/s Transmission over 50 m POF . . . 111
7.3.5 170 Mbit/s Transmission over 115 m SI-POF . . . 114

8 Equalizer Implementations for SI-POF . . . 117
8.1 Equalized Giga-Bit Transmission over SI-POF . . . 117
8.2 Fully Integrated Optical Receiver Containing an Equalizer . . . 118
8.2.1 The Integrated SI-POF Equalizer . . . 120
8.2.2 Experimental Results for Two-Stage Equalizer . . . 120
8.2.3 Experimental Results for Single-Stage Equalizer . . . 123
8.3 Summary of Integrated Optical Receivers with Integrated SI-POF Equalizer . . . 128

9 Conclusions . . . 129

References . . . 131

About the Authors . . . 137

Index . . . 139

Symbols

A_o	Open loop gain
BW	Bandwidth (Hz)
C	Channel capacity (bit/s)
C_{fb}	Feedback capacitor (F)
C_T	Total input capacitance (F)
D	Damping factor
E_b	Signal energy per bit (J)
E_g	Energy of basic symbol pulse (J)
E_{av}	Average energy (J)
E_{bav}	Average energy per bit (J)
ER	Extinction ratio
f_T	Transition frequency (Hz)
g_m	Transconductance (A/V)
G	Gain
h(z)	Channel transfer function in Z domain
$i_{n,in}$	Equivalent input noise current (A)
$v_{n,in}$	Equivalent input noise voltage (V)
I_B	Base current (A)
I_C	Collector current (A)
I_{ph}	Photo current (A)
I_{pd}	Photodiode current (A)
K_B	Boltzmann's constant (J/K)
M	Number of signal levels
NA	Numerical aperture
n_o	Refractive index
$n_{cladding}$	Cladding refractive index
n_{core}	Core refractive index
N	Average received noise power (W)
N_o	Noise power spectral density (W/Hz)
P_{AVG}	Average optical power (W)
P_{MOD}	Modulated optical power (W)

r_b	Base resistance (Ω)
R_{fb}	Feedback resistor (Ω)
R_b	Bit rate/data rate (bit/s)
R_s	Symbol rate (Baud)
S	Average received signal power (W)
S_{slop}	Laser slop efficiency
t_d	Drift time through the intrinsic zone (s)
T_s	Symbol time (s)
T_b	Bit time (s)
β	Current gain
σ_A	Acceptance angle (degree)
σ_H, σ_L	Standard deviation for binary high and low values
Q	Quality factor/Noise distance
ω_N	Characteristic angular frequency (rad/s)
ω_{GP}	Resonance gain peak angular frequency (rad/s)

Acronyms

AGC	Automatic gain control
AMGDM	Angular mode group diversity multiplexing
APC	Automatic power control
ARB/AWG	Arbitrary waveform generator
AWGN	Adaptive white Gaussian noise
BER	Bit error rate/ratio
BiCMOS	Bipolar and CMOS technology
BJT	Bipolar transistor
BPSK	Binary phase shift keying
CB	Common base
CG	Common gate
CMA	Constant modulus algorithm
CS	Common source
CW	Continues wave
DAC, D/A	Digital to analog converter
DD	Direct detection
DFE	Decision feedback equalizer
DFF	D flip-flop
DMT	Discrete multitone
DPSK	Differential phase shift keying
DR	Data rate
EMI	Electromagnetic interference
FEC	Forward error correction
FET	Field effect transistor
FFE	Feed forward equalizer
FIR	Finite impulse response
FSK	Frequency shift keying
FSM	Finite states machine
FTTH	Fiber to the home
GOF	Glass optical fiber
IC	Integrated circuit

IIR	Infinite impulse response
IM	Intensity modulation
IMD	Intermodulation distortion
IMP3	Third-order intermodulation
ICDF	Inverted-based common drain feedback
IPTV	Internet protocol television
LAN	Local area network
LA	Limiting amplifier
LED	Light emitting diode
LD	Laser diode
LMS	Least Mean Squares
MCML	MOSFET current mode logic
MLSE	Maximum likelihood sequence estimator
MII	Media independent interface
MOST	Media oriented systems transport
MOSFET	Metal/Oxide/Semiconductor field effect transistor
M-PAM	Multilevel plus amplitude modulation
M-QAM	Multilevel quadrature amplitude modulation
NRZ	Non return to zero
OEIC	Optoelectronic integrated circuit
OMA	Optical modulation amplitude
PC	Personal computer
PCB	Printed circuit board
PD	Photodiode/photo detector
PCI	Peripheral component interconnect
PIN	P-type/I/N-type semiconductor diode
PMMA	Poly-methyl methacrylate
POF	Plastic/Polymer optical fiber
PRBS	Pseudo random bit sequence
PSK	Phase shift keying
RC-LED	Resonant cavity light emitting diode
RLS	Recursive least squares
RGC	Regulated cascode
RS	Reed–Solomon code
RSSI	Received signal strength indication
SCORE	Spectral coherence restoral algorithm
SER	Symbol error rate/ratio
SI	Step index
SNR	Signal to noise ratio
STB	Set top boxes
TSPC	True single-phase clocked
TIA	Transimpedance amplifier
VCSEL	Vertical cavity surface emitting laser
VDSL	Very high speed digital subscriber line
VSG	Vector signal generator

VSA	Vector signal analyzer
VGA	Variable gain amplifier
WDM	Wave division multiplexing
WLAN	Wirelesslocal area network
ZF	Zero forcing algorithm

Chapter 1
Introduction

The increasing demand for broadband services raises the need for a high bandwidth link which should extend from the terminals to the customer's premises. In-building networks presently are using a wide range of transmission media like coaxial copper cables, twisted copper pair cables, free-space infrared links, wireless local area network (WLAN) links, etc. Each of these networks is optimized for a particular set of services; this complicates the introduction of new services and the creation of links between services (such as between video and data services). A single broadband multi-services network could provide an efficient solution to host and connect all existing and upcoming services together.

The target for data rate (DR) delivered in home could be up to 1.25 Gbit/s in case of fiber to the home (FTTH) or up to 125 Mbit/s in case of very high bit rate digital subscriber line (VDSL2) technology [1]. The home network must not represent a bottleneck for the expected evolution for services such as the introduction of high definition internet protocol television (IPTV), multi-room and multi-vision configuration and high quality video communication via the TV set. The home network can be used, for example, to share multimedia contents not necessarily delivered in real time by access network, this content can be stored in a device inside the house and used afterwards.

At present, twisted pair and coaxial cables are used as the physical medium to deliver telecom services within the customer's premises. These two transmission mediums suffer from serious shortcomings when they are considered to serve the increasing demand for broad-band services. For instance, twisted pair has a limited bandwidth and it is susceptible to electromagnetic interference (EMI). Coaxial cable offers a large bandwidth, but it poses practical problems due to its thickness and the effort required to make a reliable connection.

Moreover, the coaxial cable is not perfectly immune to EMI and has a certain attenuation. Optical fiber is extensively used for long-distance data transmission and it represents an alternative for transmission at the customer premises as well optical fiber connections offer complete immunity to EMI [2]. Glass optical fibers (GOF), however, are not suitable for use within the customer premises because of the

M. Atef and H. Zimmermann, *Optical Communication over Plastic Optical Fibers*, Springer Series in Optical Sciences 172, DOI: 10.1007/978-3-642-30388-3_1,

requirement of precise handling, and thus, the high costs involved. On the other hand, it is important to have very simple and low-cost solutions. Also the enormous capacity of the single-mode GOF is never necessary in this short distance application. The cheap Poly-Methyl-Methacrylate Plastic Optical Fiber (PMMA-POF) is an excellent candidate for implementing such a short distance network.

POF systems provide benefits compared to GOF and copper wire, which include simpler and less expensive components, operation in the visible range (the transmission windows are 530, 570, and 650 nm), greater flexibility and resilience to bending, shock and vibration, ease in handling and connecting (standard step-index POF core diameters are 1 mm compared to 8–100 μm for glass fibers), use of simple and inexpensive test equipment. Finally, POF transceivers require less power than copper transceivers. These advantages make POF very attractive for use within in-building networks [2].

The main disadvantage of the PMMA POF is its high transmission loss (150 dB/km at 650 nm and less than 90 dB/km at 530 and 570 nm) which limits the use of PMMA plastic fibers for transmitting light to less than 100 m. Most cheap commercial POFs have a uniform or step index of refraction that is the same across the core of the fiber, and step-index fibers (SI-POF) have the lowest bandwidth among multimode fibers [2]. This small bandwidth limits the maximum DR which can be transmitted through POF.

To overcome the problem of the POF's high transmission loss very sensitive receivers must be used to increase the transmitted length over PMMA POF. There are two methods to solve the SI-POF limited bandwidth problem: the first method is to use multilevel signaling like Multilevel Pulse Amplitude Modulation (M-PAM) and Quadrature Amplitude Modulation (M-QAM). Also the DR can be increased more with the limited POF bandwidth by using spectral efficient modulation techniques like Discrete Multi-Tone (DMT) [3–6]. In the second method, equalization techniques can also be used to compensate for the SI-POF limited bandwidth. Fixed or adaptive equalizers can be used for pre/post-equalization with different digital or analog equalization methods to increase the POF's bandwidth.

Until now, little effort was done to design an integrated optical receiver with a good performance for multilevel modulation. The optical receiver used with multilevel signaling must deliver a linear response over a large input optical power range. This leads to a more sophisticated design of the automatic gain control (AGC) compared to conventional binary receivers, where a significant part of the large dynamic range is achieved by a simple limiting amplifier.

To receive the multilevel optical signal we need a high linearity optical receiver with multilevel signal. The use of the conventional optical receivers (with limiting amplifier) will not be useful. The use of conventional optical receivers with limiting amplifiers will result in a distorted output signal with unequal voltage levels. This makes the signal decoding very hard or even impossible. In the design of the multilevel signaling optical receiver a linear optical receiver (no limiting amplifiers) will be considered to have equally spaced output signal voltage levels. Also a linear AGC is needed to have a constant output voltage over a wide range of input optical power. This equally spaced output signal voltage levels will ease the selection of the decision levels for signal decoding from multilevel signal to the original binary signal.

Equalization techniques can also be used to overcome the SI-POF limited bandwidth. Pre-equalization for the light source and post-equalization techniques can be used to equalize for the SI-POF small bandwidth [7–9]. Pre-equalizing of light source (peaking) lowers the light source modulation depth; this reduces the actual power per pulse compared to rectangular pulses without peaking. This is at the expense of the system power budget. A post equalizer introduces additional noise and a higher optical power is needed to achieve the required bit error rate. It follows that the use of pre- or post-equalization methods are of particular interest in systems that have adequate power reserves. Also, for a fixed equalization if the frequency response changes, as a result of different lengths of the POF or a bend in the fiber, the result will be too much or too little compensation and the bit error rate (BER) will increase. The integration of an adaptive POF equalizer with the optical receiver will reduce the system cost and enhance the performance compared to a discrete or fixed POF equalizer [10, 11].

For many applications it is desirable to integrate the photodiode (PD) with a transimpedance amplifier (TIA), POF equalizer, limiting amplifiers, and line driver into the same chip. Placing the TIA adjacent to the PD improves the performance by reducing lead capacitance and sensitivity to interference, thereby giving higher speed and lower noise. A further advantage of the integration of PD with TIA is the reduction in the external circuitry required. Hence overall cost and PCB board size can be reduced.

In this book we show how the fully integrated optical receivers fabricated with a low-cost Si process can help to achieve high transmission performance over standard PMMA SI-POF without any error correction techniques.

Chapter 2
Multilevel Signaling

2.1 Introduction

The use of multilevel signaling has applications in increasing the information bandwidth over optical fiber link, thus obviating the necessity to replace cable whose capacity is fully utilized at a given bit rate. The communication end-to-end delay can be minimized and the reduction in signal to noise ratio (SNR) caused by multilevel signaling is not significant over short distances of optical fiber. The use of multilevel signaling for light Intensity Modulation Direct-Detection (IM-DD) systems has received an attention in the literature [12–15].

By using multilevel modulation the cost and complexity are reduced by using one channel compared to binary signal over multiple channels. By using multilevel signaling over a single channel instead of multiple channels, there is no attenuation due to wavelength multiplexing at the transmitter and demultiplexing at the receiver, no multiple optics precision alignment variances, no reliability issues associated with individual channel failures and no skew management and associated delay. Wavelength Division Multiplexing (WDM) requires N lasers whereas multilevel signaling reduces the system cost by using just one similar laser.

The principle of multilevel signaling is to use a larger alphabet of M symbols to represent data, so that each symbol can represent more than one bit of data. As a result, the number of symbols that need to be transmitted is less than the number of bits (the symbol rate is less than the bit rate), hence the bandwidth is compressed. Figure 2.1 shows an example for a four-level scheme.

The bandwidth required for transmission of binary digital waveforms may be very large. For a channel of bandwidth equal to BWHz the maximum symbol rate (baud) is $R_S = 2 \cdot BW$ symbols per second. In the case of binary signaling each symbol carries one bit of information, so the information rate is limited to R_s bits per second; data rate (DR) equal to symbol rate (R_S) in binary signal case. Clearly one can increase the information rate through a channel by increasing the bandwidth or the associated symbol rate. However, if the channel bandwidth is to remain fixed, then the only option is to increase the amount of information encoded in a symbol.

M. Atef and H. Zimmermann, *Optical Communication over Plastic Optical Fibers*, Springer Series in Optical Sciences 172, DOI: 10.1007/978-3-642-30388-3_2,

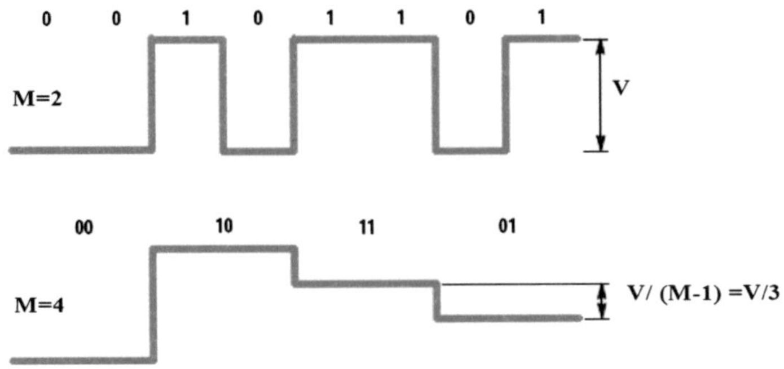

Fig. 2.1 Binary and multilevel signaling

If M is the number of distinct signal levels, then each symbol now carries $\mathrm{N} = \log_2(M)$ bits of information, and the overall data rate rises to [16]:

$$\mathrm{DR} = \mathrm{R_s} \log_2(M) \tag{2.1}$$

No additional bandwidth is required for this increase. The increased information rate comes either at the expense of added transmitter power or an increased error rate at the receiver.

The multilevel signaling scheme is more sensitive to the non-linearity and noise than the binary schemes.

2.2 Multilevel Signaling and Channel Capacity

Reducing the data bandwidth by increasing the number of signal levels comes at a price. To begin with, the modulation circuitry generating, and consequently detecting, the multilevel signal is more complex and likely consumes more power and chip area. Additionally, breaking down the available voltage into a larger number of smaller vertical segments immediately reduces the voltage margin of the received symbol by a factor of $\frac{1}{M-1}$, as shown in Fig. 2.1.

There is an extra power penalty by using M-levels compared to a binary signal [16]:

$$\text{Power Penalty} = 10 \log_{10}(M - 1)\ \text{dB} \tag{2.2}$$

Such amplitude margin loss leads to the need for increased receiver sensitivity in addition to receiver complexity to decode the incoming multilevel signal. This increases the receiver physical area and the cost.

Transition to a non-binary approach only makes sense if the penalty for doing so is less than the penalty corresponding to a standard binary approach. This is the case with transmission channel which has a roll-off higher than the power penalty. Also,

multilevel signal is used for system with high SNR to improve spectral efficiency [17]. For system with low SNR and roll-off smaller than the power penalty, binary signal is the optimal modulation format.
A comparison of the effectiveness of duobinary signaling, multilevel signal and standard binary modulation is shown in [18]. By superimposing the power spectral densities of the various modulation techniques onto the simulated for the expected channel response, it is possible to compare the bandwidth versus loss characteristics of each method. Depending on the frequency roll-off of the channel, any one of these techniques, including standard binary transmission, may be the most appropriate choice. For example, 4-PAM modulation may reduce the bandwidth requirement, but if the corresponding voltage margin degradation is too severe, it may be preferable to move to the duobinary format, in which the overall voltage is only segmented into three rather than four levels. There may even be circumstances when the higher bandwidth required by binary modulation must be accepted to achieve the optimal data eye opening for a given data rate [19].
From the previous discussion, it should be noted that such spectral efficiencies must be weighed against the cost of implementation in terms of power, chip area, and SNR penalty. In addition; such an analysis must be completed on an application and channel.
The Shannon capacity theorem has discussed the capacity of the band limited Additive White Gaussian [20–22]. The channel capacity can be calculated by:

$$C = BW \log_2\left(1 + \frac{S}{N}\right) \quad \text{[bits/s]} \quad \text{dB} \tag{2.3}$$

where BW is the Bandwidth (Hz); $S = E_b C$ the Average received signal power (W); $N = N_0 BW$ the Average noise power (W).
Increasing the signal power will mean that we can split the signal level into more levels even while ensuring low probability of error. Hence increasing the signal power will lead to more capacity. However, as the increase in capacity is a logarithmic function of power, the returns are diminishing. We can observe that increasing in channel capacity is diminishing as we keep increasing the value of power.
The second variable to play with is the bandwidth. Increasing the bandwidth has two effects:

1. More bandwidth means we can have more transmissions per second, hence higher capacity.
2. However, more bandwidth also means that there is more noise power at the receiver.

The latter effect reduces the performance, i.e. this means that increasing bandwidth alone will not lead to an increase of the capacity.

$$C = BW \log_2\left(1 + \frac{S}{N}\right) \begin{cases} S = E_b C \\ N = N_0 BW \end{cases} \tag{2.4}$$

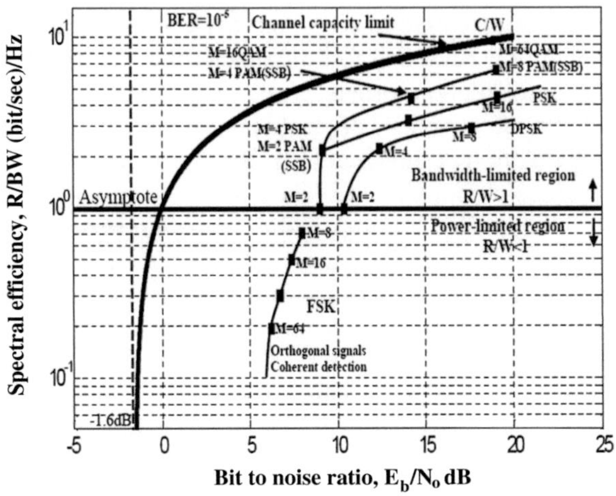

Fig. 2.2 Bandwidth efficiency-power efficiency chart for different modulation schemes [21]

$$\frac{C}{BW} = \log_2\left(1 + \frac{E_b}{N_0}\frac{C}{BW}\right) \tag{2.5}$$

$$\text{As } BW \to \infty \quad \text{or} \quad \frac{C}{BW} \to 0, \quad \text{we get:} \tag{2.6}$$

$$\frac{E_b}{N_0} \to \frac{1}{\log_2 e} = 0.693 \approx -1.6\,[\text{dB}] \tag{2.7}$$

This means that for reliable communication, we need to have $E_b/N_o > -1.6\,\text{dB}$.
From our discussion till now, we have understood that a practical communication should have a data rate DR which is lower than the capacity C, i.e. this means that increasing bandwidth alone will not lead to increase in the capacity.
Figure 2.2 shows the bandwidth efficiency-power efficiency chart for different modulation schemes. The curve divides the chart into two main regions:

(a) In the region below the curve reliable communication is possible and
(b) In the region above the curve, reliable communication is not possible.

The horizontal axis of this plot is the energy-per-bit (E_b) divided by the noise power per Hz (N_o). This ratio is called E_b/N_o and it represents the signal-to-noise ratio at the receiver in normalized form. The vertical axis is the spectral efficiency, which is the capacity divided by the bandwidth. For our discussion we think of this simply as the data rate in bits-per-second, divided by the bandwidth in Hz of the channel used by the communication system. Spectral efficiency therefore has units of bits per second per Hz. What Shannon had shown is that, it is possible to design communication systems operating with combinations of spectral efficiency and E_b/N_o below the curve, and make them work with small BER. For any system attempting to operate with spectral efficiency and E_b/N_o values above the curve, such systems cannot operate with small BER. In other words, any reliable communication system must operate below the curve. The plotted points in Fig. 2.2 are corresponding to some conventional modulation techniques. Pulse-Amplitude-Modulation (PAM), Phase Shift Keying (PSK), Differential Phase Shift Keying (DPSK), and Quadrature Amplitude-Modulation (QAM) of orders 2, 4, 8, 16, and 32 (Note that 2-QAM is really called BPSK). The Frequency-Shift-Keying (FSK) of orders 2, 4, 8, 16, 32 and 64. The E_b/N_o values correspond to BER of 10^{-5}. Figure 2.2 shows the tradeoffs among error probability, power efficiency and bandwidth efficiency. For modulations like PAM, PSK and QAM, such that when M is increased their power efficiency decreases, while their bandwidth efficiency increases. For modulations like FSK when M is increased their power efficiency increases, while their bandwidth efficiency increases.

It is customary to divide the achievable region in this chart in a bandwidth-limited region ($R_S/W > 1$) and a power-limited region ($R_S/BW < 1$). In the bandwidth-limited region, the system bandwidth is at a premium, and should be traded for power (i.e., E_b/N_o). PAM and QAM are effective modulation schemes in this region, as their bandwidth efficiency increases by increasing the number of their levels. On the other hand, FSK scheme operate in the power limited region.

It makes inefficient use of the bandwidth, but trades bandwidth for reduction in E_b/N_o necessary to achieve a given BER. The choice of modulation scheme that achieve a target BER will be guided by this chart [20–22].

The two major communication resources are the transmitted power and channel bandwidth. In many communication systems, one of these resources is more precious than the other. Hence, systems can be classified as:

1. Bandwidth-limited systems: we need to save bandwidth at the expense of power by using spectrally efficient modulation schemes (M-PAM and M-QAM). In this system, the bandwidth efficiency can be calculated by:

$$\frac{R_b}{BW} = \frac{\log_2 M}{BWT_s} = \frac{1}{BWT_b} \quad \text{[bits/s/Hz]} \tag{2.8}$$

Assuming Nyquist filtering at baseband, the required pass-band bandwidth is:

$$BW = 1/T_s = R_s \quad \text{[Hz]} \tag{2.9}$$

$$R_b/BW = \log_2 M \quad \text{[bits/s/Hz]} \tag{2.10}$$

Bandwidth efficiency increases as M increases.

2. Power-limited systems: we need to save power at the expense of bandwidth by using coding schemes like M-FSK.

$$R_b/W = \log_2 M/M \quad \text{[bits/s/Hz]} \tag{2.11}$$

Bandwidth efficiency decreases as M increases.

From the above discussion SSB M-PAM and M^2-QAM are the best choice for bandwidth limited channels like PMMA SI-POF channel. One important advantage of M-PAM modulation scheme compared to the M^2-QAM modulation is that transmitter and receiver architectures are easier to design. Also it is shown that baseband M-PAM offers a 1.5 dB sensitivity improvement compared to M^2-QAM due to its lower symbol distance [23].

2.3 Optimum Number of PAM Levels

Now the question of how many levels of PAM comes, 4, 8 or more. We need to consider several factors here. If the levels are more, the spacing between the adjacent levels will be less. And less spacing means more susceptibility to the noise and ISI. Moreover, the sensitivity of the receiver has to be very high. Well, this limited receiver sensitivity might be circumvented if the spacing is made larger by using a large swing at the transmitter end. But, for maintaining the same spacing, more levels mean larger swing at the transmitter end. However, the transmitter has also limited output swing. It should be noted that the number of levels must be weighed against the cost of implementation multilevel transmitter and receiver in terms of power, chip area and SNR penalty. In addition; such an analysis must be completed on an application and channel. Depending on the frequency roll-off of the channel, any number of levels can be the most appropriate choice.

We next outline explicit examples whereby 4-PAM signaling might become a plausible option for doubling the baud rate for limited bandwidth channel.

First, by substituting in Eq. (2.2) the SNR overhead degradation for deploying 4-PAM versus binary signaling at same baud rate is:

$$SNR_{degradation} = 20\log(2^2 - 1) = 9.6\,\text{dB} \tag{2.12}$$

The laser nonlinearity has a higher effect on the 4-PAM performance than on binary signal. The laser nonlinearity can be compensated to decrease its bad effect on the 4-PAM signal but there is still a penalty coming from the laser nonlinearity. From our measurements in Chap. 5, it is nearly 1.5 dB.

To deliver the same data throughput, the baud rate of binary signaling needs to be doubled resulting in a bandwidth spectrum of the signal twice as wide as 4-PAM. For 4-PAM signaling a receiver with half the bandwidth of the binary receiver can be used. A noise analysis for our integrated optical receiver using SPECTRE results in that the optical receiver input noise power will be decreased by 3 dB by using receiver having half the binary receiver bandwidth. The output electrical noise will be decreased by 6 dB. The SNR degradation can be written as:

$$SNR_{degradation} = 9.6\,\text{dB} + 1.5\,\text{dB} - 6\,\text{dB} = 5.1\,\text{dB} \tag{2.13}$$

$$SNR_{degradation} = \log(2) \cdot SNR_{slope} \tag{2.14}$$

where SNR_{slope} is the bandwidth roll off (per decade) of a transmitter's response. From (2.13) and (2.14) the 4-PAM becomes advantageous over binary when the following conditions develop:

$$SNR_{slope} > 17\,\text{dB/Decade} = 5.1\,\text{dB/Octav} \tag{2.15}$$

When driving the data rate of the fiber channel beyond the small-signal characteristic's 3 dB point, the roll-off exceeds 5.1 dB/octave, under this condition such a fiber channel could advantageously use 4-PAM to double the data rate instead of signaling with binary.

2.4 M-PAM Symbol Error Rate

A detailed analysis for symbol error rate (SER) can be found at [20, 22]. Consider M amplitude levels centered on zero, with $M = 2^l$. If the M-PAM signals are represented geometrically as M one dimensional signal point values:

$$S_m = \sqrt{0.5E_g A_m}, \quad m = 1, 2, \ldots, M \tag{2.16}$$

where E_g is the energy of the basic signal pulse. The amplitude values may be expressed as:

$$A_m = (2m - 1 - M)d, \quad m = 1, 2, \ldots, M \tag{2.17}$$

where the Euclidean distance between adjacent signal points is $d\sqrt{2E_g}$. The average symbol energy E_{av} can be calculated from:

$$E_{av} = \frac{1}{M}\sum_{m=1}^{m} E_m = \frac{d^2 E_g}{2M}\sum_{m=1}^{m}(2m - 1 - M)^2$$

$$E_{av} = \frac{M^2 - 1}{6} d^2 E_g \tag{2.18}$$

Placing the threshold levels as shown in Fig. 2.3 helps in evaluating the probability of error. We note that if the Mth amplitude level is transmitted, the demodulation output will be:

$$r = S_m + n = \sqrt{0.5E_g}\,A_m + n \tag{2.19}$$

Where the noise variable n has zero mean and variance $\sigma^2 = 2N_o$ and all amplitude levels are equally likely a priori. The average probability of a symbol error is simply the probability that the noise variable n exceeds in magnitude one-half of the distance between levels. Thus we have

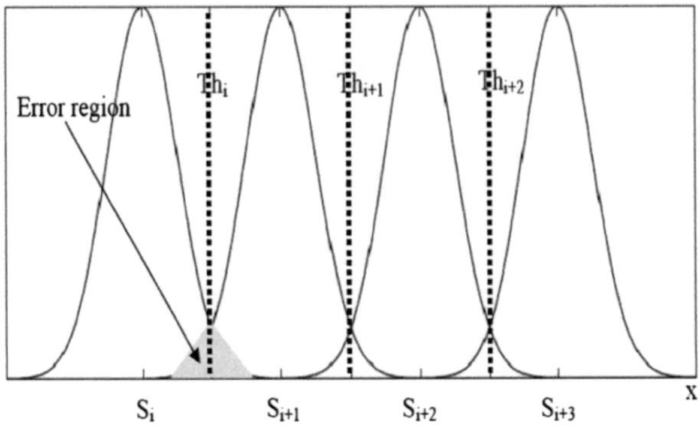

Fig. 2.3 Power density function for M-PAM signal

$$P_M = \frac{M-1}{M} P(|r - S_m| > d\sqrt{0.5E_g})$$
$$= \frac{M-1}{M}\frac{2}{\sqrt{\pi N_o}} \int_{d\sqrt{0.5E_g}}^{\infty} e^{\frac{-x^2}{N_o}} dx \tag{2.20}$$
$$\text{SER} = \frac{M-1}{M}\frac{2}{\sqrt{\pi}} \int_{\sqrt{d^2E_g/N_o}}^{\infty} e^{\frac{-x^2}{2}} dx$$
$$\text{SER} = P_M = \frac{2(M-1)}{M}\text{Q}\left(\sqrt{\frac{d^2E_g}{N_o}}\right) \quad \text{where} \quad \text{Q(x)} = 0.5\,\text{erfc}\left(\frac{x}{\sqrt{2}}\right).$$

The SER can be written in terms of average symbol energy Eq. (2.18):

$$\text{SER} = \frac{2(M-1)}{M}\text{Q}\left(\sqrt{\frac{6E_{av}}{(M^2-1)N_o}}\right) \tag{2.21}$$

It is customary to plot the probability of symbol error (SER) to use SNR per bit as the basic parameter, $E_{av} = Log_2(M)E_{bav}$ where E_{bav} is the average bit energy:

$$\text{SER} = \frac{2(M-1)}{M}Q\left(\sqrt{\frac{6Log_2(M)E_{bav}}{(M^2-1)N_o}}\right) \tag{2.22}$$

BER can be obtained by dividing SER by the number of bits per symbol $Log_2(M)$. The SER presented in Eq. (2.22) is illustrated at Fig. 2.4 as function of average signal to noise ratio per bit $\left(\frac{E_{bav}}{N_o}\right)$.

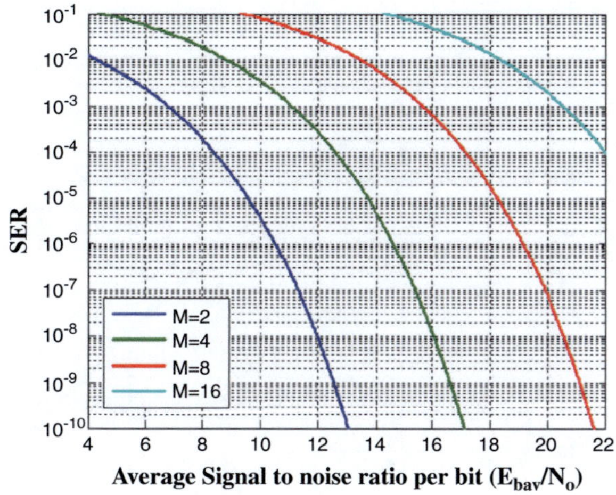

Fig. 2.4 Symbol error rate for M-PAM signal

2.5 Bit Error Rate for Binary Signals

For binary signals the number of levels M is equal to 2. The quality factor Q which measures the quality of the signal is defined by:

$$Q = \frac{P_H - P_L}{\sigma_H + \sigma_L} \tag{2.23}$$

where P_H and P_L are the signal average power for high and low value, respectively, and σ_H and σ_L are their standard deviations.
By using M = 2 and substituting Eq. (2.23) into Eq. (2.22) plus considering $\sigma^2 = 2N_o$ and the function $Q(x) = 0.5\,\mathrm{erfc}\left(\frac{x}{\sqrt{2}}\right)$, the BER for a binary signal can be calculated by:

$$\mathrm{BER} = 0.5 \cdot \mathrm{erfc}\left(\frac{Q}{\sqrt{2}}\right). \tag{2.24}$$

The BER as function of the quality factor Q (noise distance) is plotted in Fig. 2.5. Communication analyzers or digital sampling oscilloscopes can be used to determine the quality factor Q form an eye diagram according to

$$Q = \frac{V_H - V_L}{\sigma_H + \sigma_L} \tag{2.25}$$

where V_H and V_L are the signal average voltage for high and low values, respectively, and σ_H and σ_L are their standard deviations.

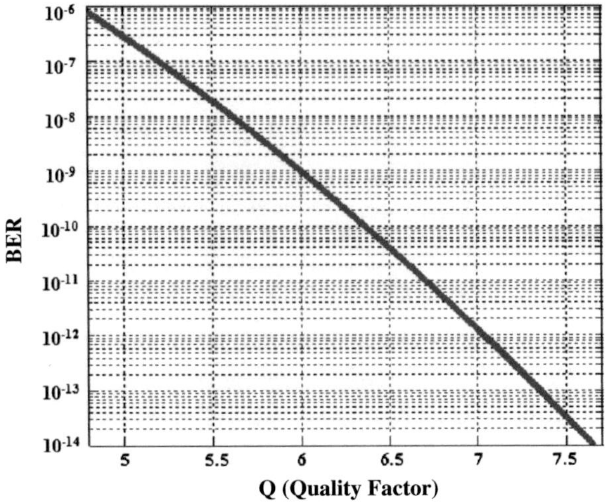

Fig. 2.5 Bit error rate as a function of the quality factor

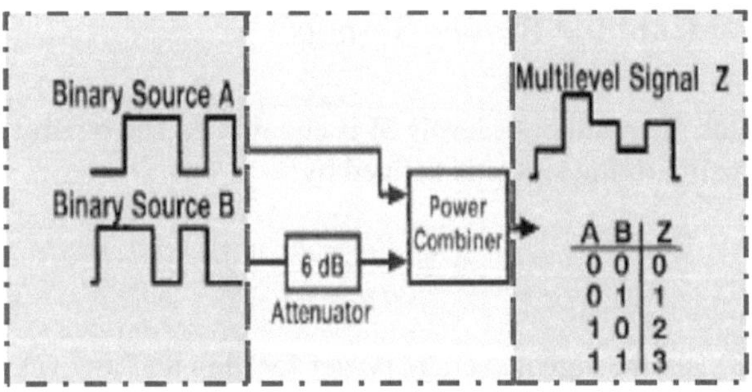

Fig. 2.6 Block diagram for multilevel signal generator using two input bits and a power combiner [14]

2.6 M-PAM Generation and Decoding

2.6.1 M-PAM Generation

Generating a 4-PAM electrical signal from two binary signals at Gigabit rates can be accomplished using the circuit shown in Fig. 2.6. There are two binary voltage sources, the first binary voltage source A is directly applied to the power combiner. The second binary voltage source B is attenuated to the half then applied to the power combiner. The binary signal B can be a delayed version from binary signal A. The output of the power combiner is the 4 levels signal Z [14].

We used this method to generate the 4-PAM signal in our first measurements. The eye diagram of the generated 4-level signal is illustrated in Fig. 2.7.

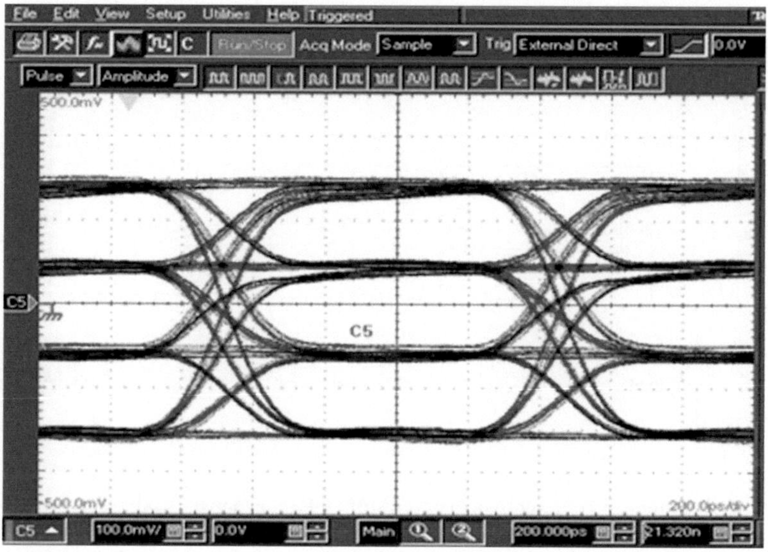

Fig. 2.7 Four level signal generated by a 6 dB attenuator and power combiner

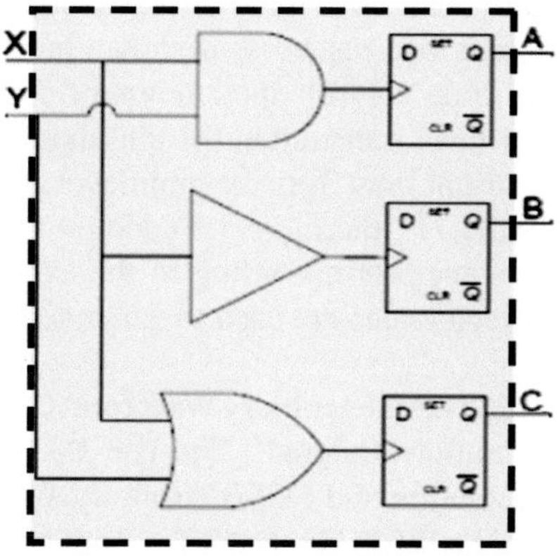

Input X Y	Output A B C
00	000
01	001
10	011
11	111

Fig. 2.8 MCML encoder gate and thermometer code to generate 4-level signal [24]

The block diagram of the multilevel transmitter by another technique is presented in Fig. 2.8. The multilevel transmitter consists of a MCML encoder which converts input data to thermometer code, a D Flip-Flop (DFF) for synchronization, a multilevel generator with a thermometer-coded current-steering Digital-to-Analog Converter (DAC) to generate the different level voltage swings, and a laser driver [24].

In order to generate 4-level signals, the information binary data (X, Y) has to be converted to thermometer code (A, B, C), in which the number of ones are determined by the corresponding input combination, shown in Fig. 2.8. The 4-level system is

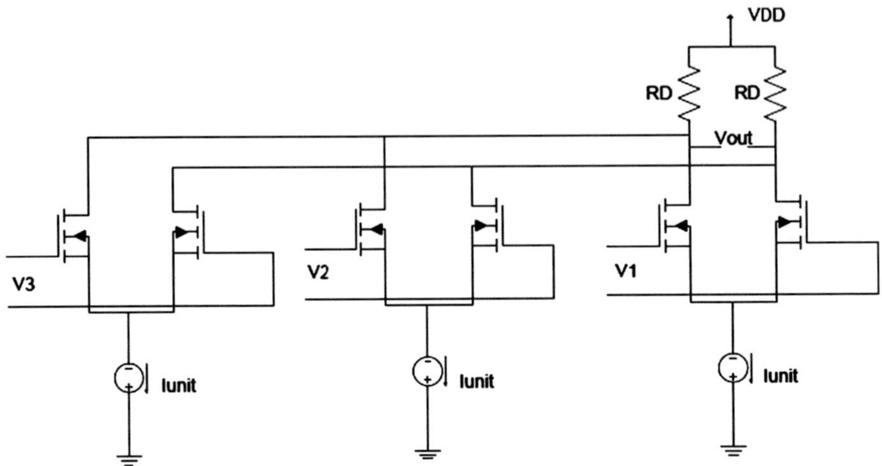

Fig. 2.9 A thermometer coded current-steering DAC 4-level generator [24]

represented by A = X · Y, B = X, and C = X + Y. Generally, with N binary bits, we have M = 2N − 1 thermometer-coded bits.

In the 4-level system, the multilevel generator consists of three current sources, three differential pairs as switches, and two passive load resistors, as shown in Fig. 2.9. The number of one bit in the thermometer code controls the amount of current in the passive load, RD. For instance, in the 4-level transmitter, the 000 thermometer code bit sequence does not allow any unit current flow from the multilevel generator to the load, thus V_{out} will be VDD. Also, the 111 thermometer code bit sequence results in an output voltage of the 4-level generator according to the expression: $V_{out} = VDD - 3 \cdot Iunit \cdot RD$. Many other techniques are used to generate M-PAM signal [13, 26, 27].

In our M-PAM measurement, the DA11000 PCI based Arbitrary Waveform Generator (ARB) is also used to generate Gigabit multilevel signal [25]. The DA11000 is one 50 ohm SMA output channel, 12-bit resolution and 1 GS/s Arbitrary Waveform Generator on a single mid-sized PCI card. The DA11000 has TTL input triggering capability that allows a segment or segments of data to be output only after a trigger is present. The block diagram and the PCB of the ARB are illustrated in Figs. 2.10 and 2.11.

The ARB can be programmed by C++ or MATLAB to generate the required multilevel signal or even more complex signals. The eye diagram for the generated 4 level signal using the ARB is shown in Fig. 2.12.

For both M-PAM and M-QAM generation at symbol rate 1.25 Gbaud we use a two orthogonal channel multilevel generator from SYMPULS [26].

The multilevel pattern generator 1250-16 PAT from SYMPULS is a special test generator to drive the laser source by multilevel signal. The output signals have a multilevel code, which can give states up to 16 levels. Each code element is,

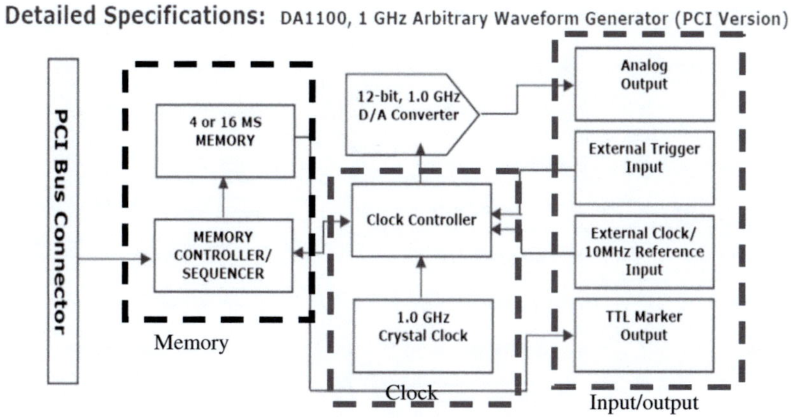

Fig. 2.10 Block diagram of PCI based Arbitrary Waveform Generator [25]

Fig. 2.11 The PCB of PCI based Arbitrary Waveform Generator [25]

therefore four Bit encoded. For operation it needs an external reference clock signal in the range from 20 MHz up to 2.5 GHz. It can repeat rates between 10 MHz and 1.25 GHz. The 32 Mbit internal memory allows the generation of freely programmable patterns of up to 4,194,304 data characters. On the front panel outputs are two independent data signals (A multilevel and multilevel B), quadrature-clock signals (90° phase shift) and a selectable trigger output for. Figure 2.13 illustrates the generated 8-level eye diagrams for the two multilevel outputs at symbol rate 1.25 Gbaud/channel.

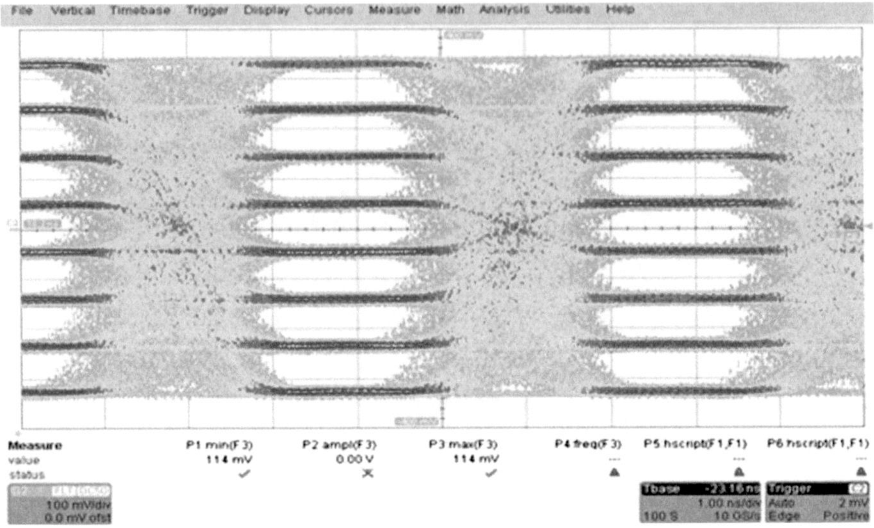

Fig. 2.12 The eye diagram for the generated 8-level signal using the PCI based Arbitrary Waveform Generator

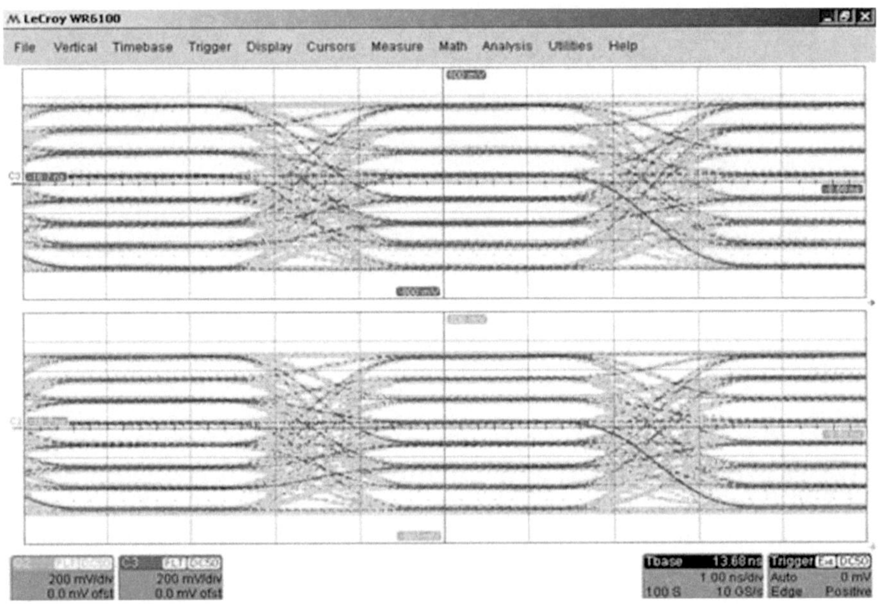

Fig. 2.13 The 8 level eye diagram for the two multilevel outputs generated by the SYMPULS multilevel generator

D	A	B	C	D1	D2
0	0	0	0	0	0
1	0	0	1	0	1
2	0	1	1	1	0
3	1	1	1	1	1

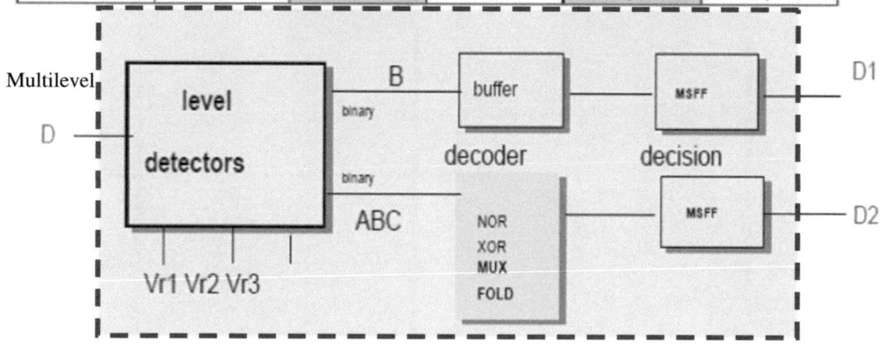

Fig. 2.14 4-Level signaling decoder

2.6.2 M-PAM Decoding

After the generation of 4-level signals which is used to drive the laser source, the optical signal is detected by an optical receiver and converted to an electrical signal. A decoding circuit is needed to decode the electrical 4-level signal to two binary signals.

Reference [27] has analyzed the architectures for a multilevel decoder-decision circuit used for 4-level. Three differential pairs compare the signal D with 3 reference voltages, giving three binary signals (labeled A, B and C), according to the truth table of Fig. 2.14.

Another decoder is shown in Fig. 2.14. This is the conventional 4-PAM decoder introduced in [12]. The implemented conventional 4-PAM decoder consists of basically three comparators followed by a 3×2 decoder (Fig. 2.15).

2.7 DC Balance Code

Most transmission systems do not pass the DC-component, a situation that is usually called as “AC-coupling”. The AWG used to generate the multilevel signal in this work is AC-coupled and has a lower cut-off frequency. Also the output of the optical receiver is AC-coupled to the external processing units. In a standard binary NRZ transmission system, AC-coupling generates the unwanted baseline wander. This wander in the resulting signal is due to the removal of the signal DC components below the low cut-off frequency. Baseline wander is particularly evident for long strings of “1” or “0”. One of the most commonly used techniques, being the one dictated by all the optical versions of the Ethernet standard, is the 8B/10B coding, described in details in [28]. The 8B/10B code maps each possible 8 bit sequence into

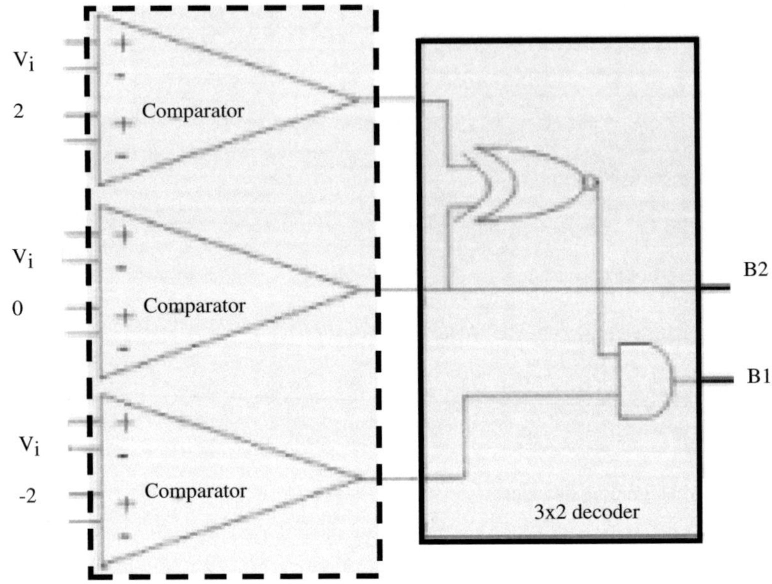

Fig. 2.15 Schematic of the conventional 4-PAM decoder [12]

a new 10 bit sequence, using a mapping that is characterized by a good DC-balance. We note that 8B/10B, besides solving the AC-coupling problem, also eases clock recovery since, again, and it avoids long sequence of bits without transitions. In case of multilevel transmission, such as in M-PAM, the problem is basically the same, so it is possible to figure out a coding somehow related to the original idea exposed in [28]. Figures 2.16 and 2.17 illustrate the effect of AC-coupling and DC-balance coding on the transmitted signal.

2.8 Multilevel Signaling Related Work

Multilevel signaling was used to extend the dispersion limited transmission distance of IM-DD fiber optic communication systems operating at 1550 nm over standard fiber over long distances. Unlike the binary schemes, it requires all logic devices to operate at only half the aggregate bit rate, addressing the speed limitation of electronic components [14, 29].

Reference [30] presents the initial steps in the development of a point-to-point IR wireless link capable of providing data rates up to 1 Gb/s over short distances. Such a link would provide a useful means for connecting portable devices to network and video data stores as well as to better user interface peripherals. Reference [30] has discussed the need of multilevel modulation schemes for free-space optical

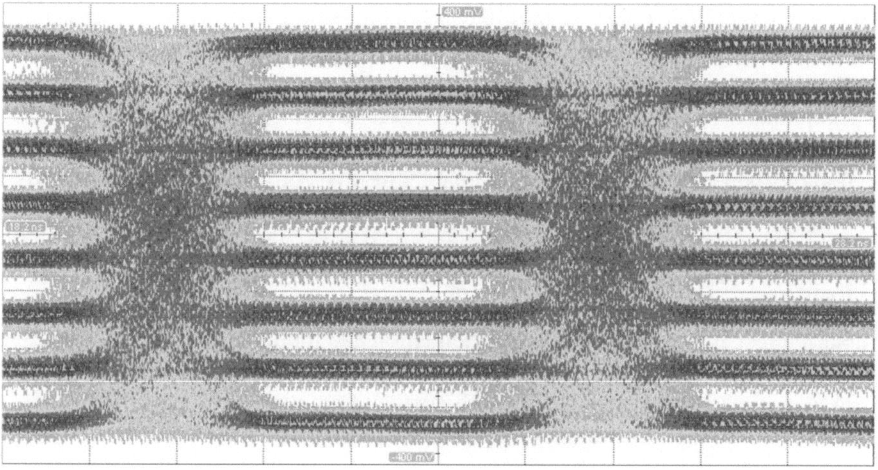

Fig. 2.16 Effect of AC-coupling on the transmitted 8-PAM signal at symbol rate 250 Mbaud and PRBS length = $2^{15} - 1$

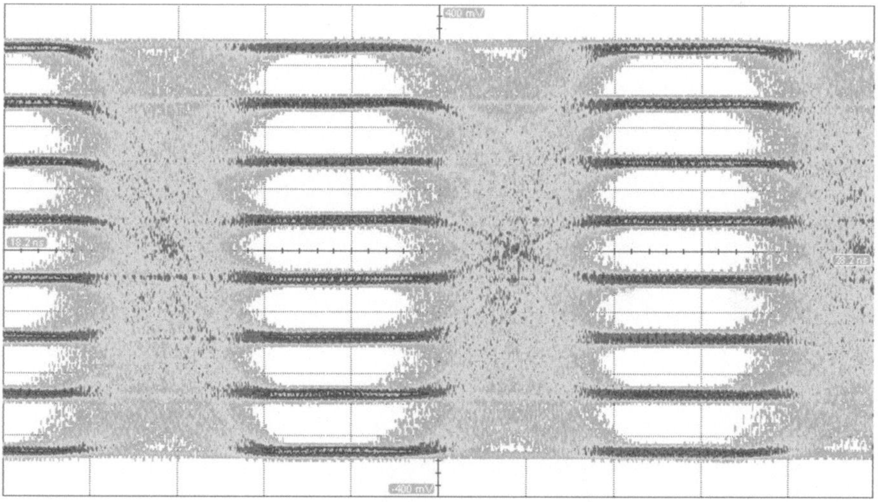

Fig. 2.17 8-PAM signal at symbol rate 250 Mbaud and PRBS length = $2^{15} - 1$ after DC-balance coding

communications. The advantages of binary modulation schemes are outweighed by their poor bandwidth efficiencies.

In the light of rapid growing interest in multimedia communications, there is a particular important application of multilevel signaling method in Gigabit Ethernet (1000Base-T), which combines 5-PAM with the use of parallel differential signal lines to increase the symbol rates while retaining a low line rate. For

devicemanufacturers, however, multilevel signaling poses new challenges for achieving reliable, high-throughput testing needed to deliver required levels of quality and production volume [13].
A comprehensive investigation of Vertical Cavity Surface Emitting Laser (VCSEL) links operating under 4-PAM signaling using commercially available 0.13 μm CMOS technology was presented in [15]. This enables the first complete comparison between PAM-4 and binary optical signaling. 4-PAM signaling is demonstrated to overcome the rise time limitations in the transmitter as well as the potential modal noise in the multi-mode fiber link that otherwise degrades binary signaling.
The M-PAM modulation was used with PMMA SI-POF to overcome the physical impairments generated by POF high attenuation and modal dispersion, which result in low power margins and very small available bandwidth [4, 31–33]. The 4- and 8-PAM modulations in conjunction with post- and pre-equalization techniques were used to increase the transmission length of the large-core PMMA SI-POF.

Chapter 3
Equalization Techniques

The term equalization can be used to describe any circuit or signal processing operation that minimizes ISI. The purpose of an equalizer is to reduce ISI as much as possible to minimize the probability of wrong decisions. The ideal response requires an equalizer that responds as the inverse of the channel. The ideal response is typically not practical; it needs more components, area and complex design. The cost and power constraints are the limits for such equalizers. The equalization target is to reduce distortion and BER to acceptable levels. The equalizer can be made of discrete components on PCB, integrated using silicon technologies like CMOS or BiCMOS or built into cables or connectors.
The equalizers can be classified as:

1. Linear equalizers and non-linear equalizers: if the equalizer does not have a feedback path, the equalization is linear. If there is a feedback path, the equalization is nonlinear like Decision Feedback Equalizer (DFE) and Maximum Likelihood Sequence Estimator (MLSE). Linear equalization techniques are simple to implement, but greatly amplify noise power because they work by inverting the channel frequency response. Non-Linear Equalization techniques are more complex to implement, but have much less noise enlargement than linear equalizers. The linear equalization is faster than the nonlinear equalization but the nonlinear equalizers can achieve a better equalization on the expense of complexity and loss of speed.
2. Pre-equalizers and post-equalizers: the equalizer may be placed anywhere in the channel, at the transmitter side (pre-equalizer), receiver side (post-equalizer), or both sides.
   Pre-equalization adjusts the magnitude of the transmitter output by adding certain peaking based on the channel response. This reduces the maximum swing, but produces an open eye.
3. Passive equalizers and active equalizers: The passive equalizer introduces only attenuation at certain frequencies. The active equalizers can attenuate or amplify certain frequencies to achieve the required equalization.

M. Atef and H. Zimmermann, *Optical Communication over Plastic Optical Fibers*, Springer Series in Optical Sciences 172, DOI: 10.1007/978-3-642-30388-3_3,

4. Fixed equalizers and adaptive equalizers: the fixed equalizer has a fixed coefficient suited for a time invariant channel. The adaptive equalizer coefficient can be adapted to equalize for a time variant channel.

A combination between the above equalization types can be made to get the one which fits the required specifications best.

3.1 Passive Equalizer

The SI-POF channel frequency response has a low bandwidth characteristic, so the required equalizer has to respond like a high pass filter.
The POF passive equalizer is a high pass filter where the low frequency components are attenuated like in a passive R-C filter (see Fig. 3.1).

$$\mathrm{A}(\omega) = \frac{z_3}{R_2 + z_2 + z_3} \tag{3.1}$$

$$z_2(\omega) = \frac{R_2}{1 + J\omega R_2 C_2}, \quad z_3(\omega) = \frac{R_3}{1 + J\omega R_3 C_3} \tag{3.2}$$

The branch C_2 // R_2 introduces a zero to the equalizer transfer function and the branch C_3 // R_3 introduces a pole. The resistance R_2 introduces attenuation to the low frequency signal. By controlling the position of the zero and pole as well as the attenuation, the equalization can be done.

3.2 Active Equalizer

Passive filters attenuate the signal and they do not have appropriate input and output impedances that the active filters have. Active filters have an operating frequency range and power range where they can be used successfully to achieve better performance than passive filters.
The low frequency components are attenuated and the high frequency components can be amplified by using an active high pass filter.
In the next sections different types of active high-pass filters will be introduced like source degeneration and inductive load amplifiers.

3.2.1 Source Degeneration

A common source amplifier with a degenerated source as an active high-pass filter is shown in Fig. 3.2. Low-frequency signals cannot pass through the branch consisting

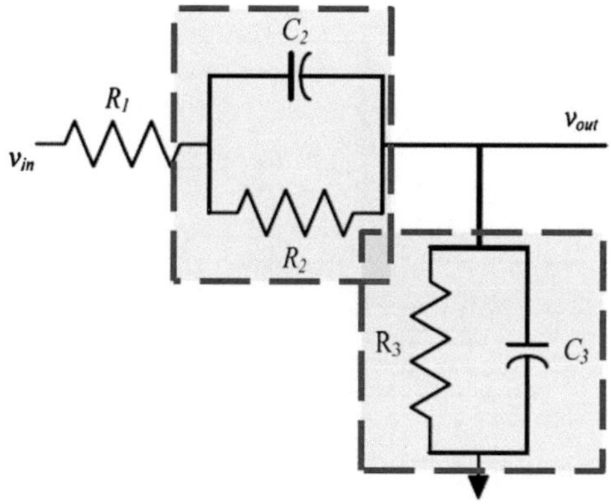

Fig. 3.1 Passive R-C equalizer

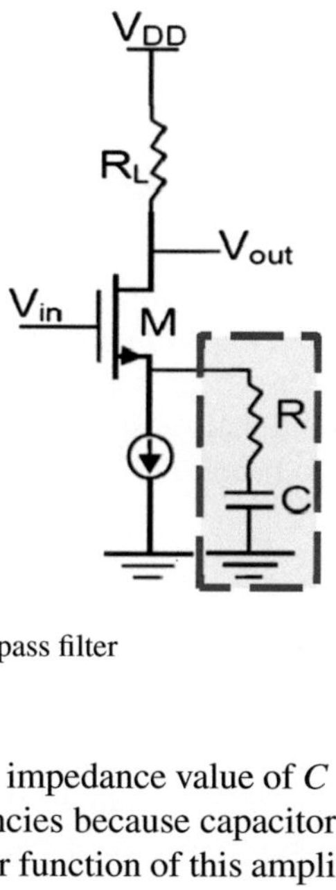

Fig. 3.2 Source degeneration high pass filter

of R and C because of the high impedance value of C at low frequencies. Signal can pass through R at high frequencies because capacitor C will have a low impedance value. The voltage gain transfer function of this amplifier has a high-pass filter characteristic:

$$A(\omega) = \frac{J\omega\, R_L C}{1 + J\omega\, C\left(R + \frac{1}{g_m}\right)} \tag{3.3}$$

$$A(\infty) = \frac{g_m R_L}{g_m R + 1}, \quad \omega_p = \frac{g_m}{C(1 + g_m R)} \tag{3.4}$$

where $A(\infty)$ the high-frequency gain and ω_p is the corner frequency of the high-pass filter.

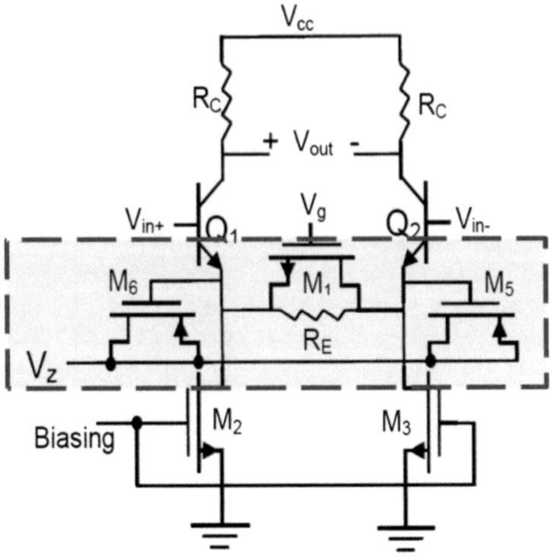

Fig. 3.3 Differential source degeneration adjustable equalizer

3.2.2 Differential Source Degeneration Adjustable Equalizer

The equalizer in Fig. 3.3 has two control inputs, the first control signal (V_z) controls the high-pass filter corner frequency and the second control signal (V_g) controls the low-frequency gain. The high cut-off frequency is controlled by varying the MOS capacitors M_5 and M_6. The additional passive resistor R_E is used to give a proper minimum low-frequency gain to prevent an undesired too low low-frequency gain when M_1 is off.

3.2.3 Inductive Load Amplifier

A common-source amplifier with an inductive load is shown in Fig. 3.4a. The equivalent impedance of the inductor *L* in shunt with the resistor *R* results in a low voltage gain at low frequencies and a high voltage gain at high frequencies. The voltage gain transfer function is [34]:

$$A(\omega) = \frac{J\,\omega\,\mathrm{L}g_m}{1 + J\omega\,\frac{\mathrm{L}}{\mathrm{R}}} \tag{3.5}$$

$$A(\infty) = g_m R, \quad \omega_p = \frac{R}{\mathrm{L}} \tag{3.6}$$

where $A(\infty)$ is the high frequency gain and ω_p is the corner frequency of the high pass filter.

An integrated inductor occupies a large chip area, which increases the costs. It suffers from a low quality factor due to the presence of a highly doped substrate [35].

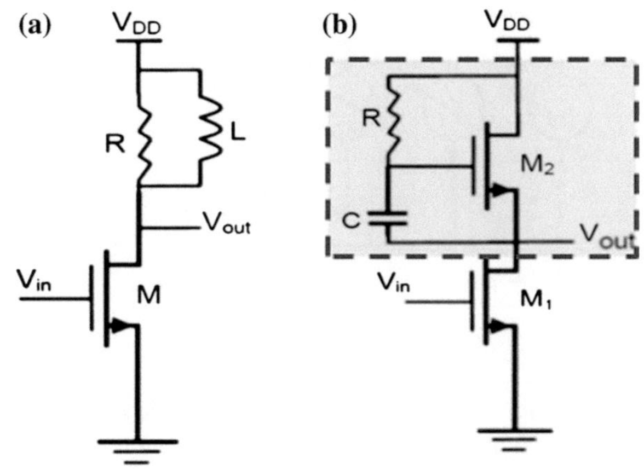

Fig. 3.4 Inductive load amplifier high-pass filter with (**a**) passive inductor (**b**) active inductor

The inductor can be replaced by an active one having the same behavior by means of devices that are easier to integrate. A common-source amplifier with active inductor is shown in Fig. 3.4b. The MOSFET M_2 makes a path for the low frequency signal where the R-C branch has high impedance due to C. At high frequencies the branch R-C creates a second signal path in parallel to M_2 and most of the signal current flows through the low impedance path R-C. At high frequencies, the load impedance is determined by R. If R is lower than $1/g_{m2}$ of M_2, the voltage gain at high frequencies is significantly higher than at low frequencies. The voltage gain transfer function of the circuit in Fig. 3.4b is given by the following relation:

$$\mathrm{A}(\omega) = \frac{g_{m1}}{g_{m2}} \cdot \frac{1 + J\omega\,\mathrm{RC}}{1 + J\frac{\omega\,\mathrm{C}}{g_{m2}}} \tag{3.7}$$

$$\mathrm{A}(\infty) = g_{m1}\mathrm{R}, \quad \omega_p = \frac{g_{m2}}{\omega\,\mathrm{C}} \tag{3.8}$$

where $\mathrm{A}(\infty)$ the high-frequency is gain and ω_p is the corner frequency of the high-pass filter.

3.3 Filter Implementation Using Transversal and Lattice Structures

The choice of filter implementation method with transversal or lattice structure depends upon the application. If we know the number of coefficients in the transversal filter or, equivalently, the number of lattice stages, then the transversal-filter implementation is the most computationally efficient one. However, the transversal-filter implementation does not generate the low-order optimal residuals that are available in

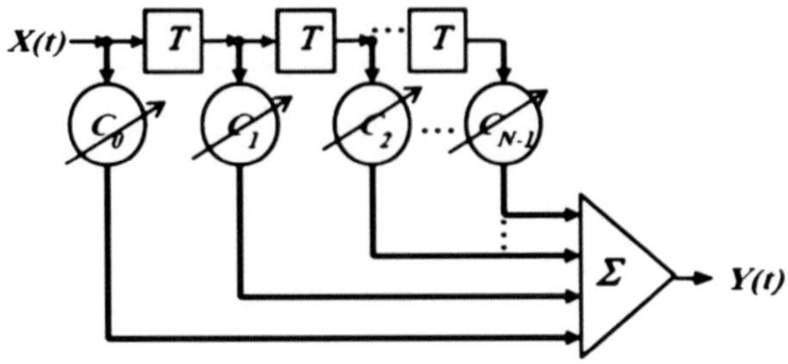

Fig. 3.5 Continuous time FIR filter implementation

the lattice implementation. In parallel processing applications, the lattice also appears to have a slight structural advantage in terms of local interconnections, though the increased computational effort somewhat offsets this advantage; on the other hand, the transversal filter approach has a substantial structural advantage when a single computational unit is time-shared for the application [36].
The FIR (Finite Impulse Response) filters use only current and past input samples and none of the filters previous output samples, to obtain the current output value. The FIR filter transfer function has only zeros. The FIR filter can be realized in many structures like the transversal and lattice implementation. A transversal FIR filter is realized by a tapped delay line Fig. 3.5. The delay line stores the past input values. The output from each tap is summed to generate the filter output. In comparison with FIR filter, the Infinite Impulse Response (IIR) filter is an efficient way of achieving a certain performance characteristics with a given filter order. This is because the IIR filter incorporates feedback and is capable of realizing both poles and zeros of a system transfer function, whereas the FIR filter can only realize zeros. Furthermore, because of the recursive structure, the IIR filter can be efficiently implemented with fewer cycles on microcontrollers and DSPs. So, IIR filters are often used in embedded applications. The problem with IIR types in analog implementations is the feedback speed. They could be done with sample and holds in the feedback loop to get the sampling effect. However this would not be as cost effective as the classical analog filter types. Analog IIR filters need more component parts, silicon area, and power consumption.

3.4 Linear Equalization

3.4.1 Continuous Time FIR Filter Implementation

The analog FIR filter is implemented using variable gain blocks and delay elements, Fig. 3.5. The passive delay line is easy to design but it requires a large CMOS area and suffers from limited design freedom because of characteristic impedance matching requirements, thus limiting the choice of usable on-chip inductors. For the above mentioned reasons, an active delay line version is better to use for integrated circuits.

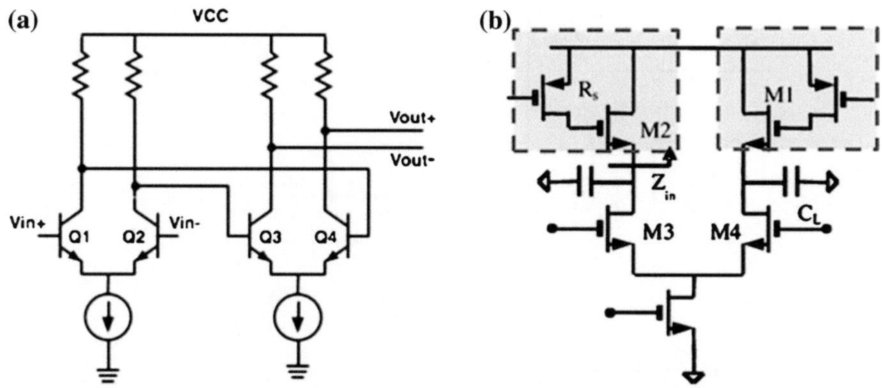

Fig. 3.6 Active delay line implementation using (**a**) two cascaded bipolar differential amplifiers [37] and (**b**) a MOSFET differential amplifier with active inductor loads [38]

The active delay line is more compact and has lower power consumption. Another important benefit of the active delay line is that it does not introduce DC voltage drops across each stage, and the overall bandwidth is not drastically degraded by additional delay stages. An example of an active delay line implementation is shown in Fig. 3.6a. There is a trade-off between the delay and bandwidth. A higher delay-bandwidth product is possible if a peaking effect is applied in the frequency domain transfer function, but the accumulation of this peaking effect in the filter can pose a problem at the summing node by causing waveform distortion [37, 38].
To integrate further and implement a varying time delay, the source follower as a load replaces the series combination of an on-chip inductor and resistor. The active delay-line cell is shown in Fig. 3.6b [38].
The overall voltage gain for a differential pair in the unit delay cell is [38]:

$$A(\omega) = g_{m3} \cdot (Z_{in}//C_L) \tag{3.9}$$

$$Z_{in} = (S \cdot R_s \cdot C_{gs2} + 1)/(S \cdot C_{gs2} + g_{m2}) \tag{3.10}$$

Then the overall voltage gain for a differential pair in the unit delay cell can be calculated by:

$$A(\omega) = \frac{g_{m3} \cdot (1 + J\omega\, R_s C_{gs2})}{g_{m2} + J\omega \cdot (C_{gs2} + C_L) - \omega^2\, C_{gs2} C_L R_s} \tag{3.11}$$

where R_s is the turn on resistance of M_1. There is a zero ω_z. By varying the value of R_s the zero location can be controlled.

$$\omega_z = \frac{1}{R_s C_{gs2}} \tag{3.12}$$

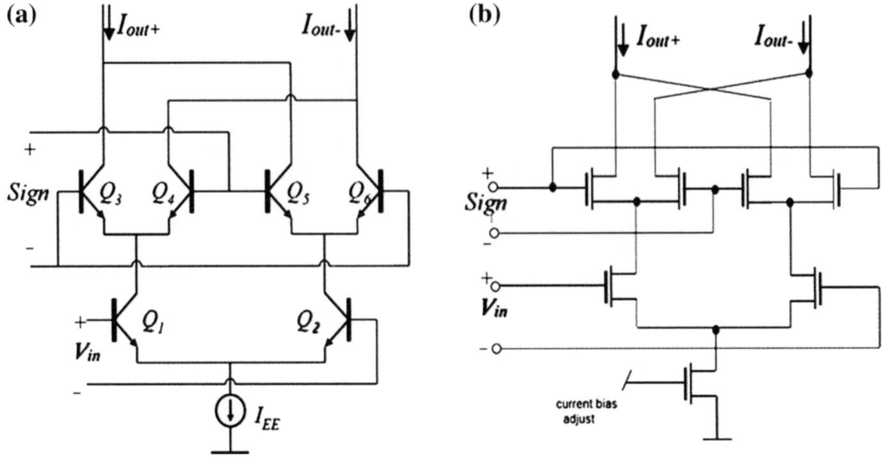

Fig. 3.7 Gilbert cell as gain stage: **a** BJT implementation, **b** NMOS implementation

The Gilbert cell, shown in Fig. 3.7a, is used as a gain stage for various tap gains. The tail current I_{EE} of a differential pair controls the tap gains ($C_0, C_1, \ldots, C_n$), while two control signals determine the polarity of the gain. The input signal is fed into Q_1 and Q_2. The upper differential pairs decide the sign of the gain depending on which pair is activated [37]. A MOSFET version from the Gilbert cell is shown in Fig. 3.7b. The Gilbert cell acts as a voltage to current converter so that the filter response can be combined from the tap at the summing node.

The simplest approach to implement the summing stage is to tie all the output currents from the multipliers together and then convert them into a voltage through a load resistor R_L. The bandwidth of the summing stage will be limited due to the large value of R_L to achieve the required gain. To improve the bandwidth, a low resistance is needed at the summing node, where the parasitic capacitance is expected to be large due to the contribution of all the multipliers. The bandwidth can be increased by decreasing R_L, but the gain will be decreased and the signal-to-noise ratio will be worse.

To increase the bandwidth of the summing circuit and have enough gain, a transimpedance load is used, as shown in Fig. 3.8. Resistors R_1 provide a feedback path for transistors M_3 to implement a TIA. Resistors R_2 bias the gate of transistors M_4 and M_5. The equivalent input resistance is given by $1/g_{m3}$ and the transimpedance gain is given by $-R_1$ for $R_2 \gg R_1 \gg 1/g_{m3}$ [39].

3.4.2 Discrete Time FIR Filter Implementation

Discrete time equalizers use FIR filters, Fig. 3.9. The input stream (x_k) propagates through a series of delay units. The signals taken between the delay units are

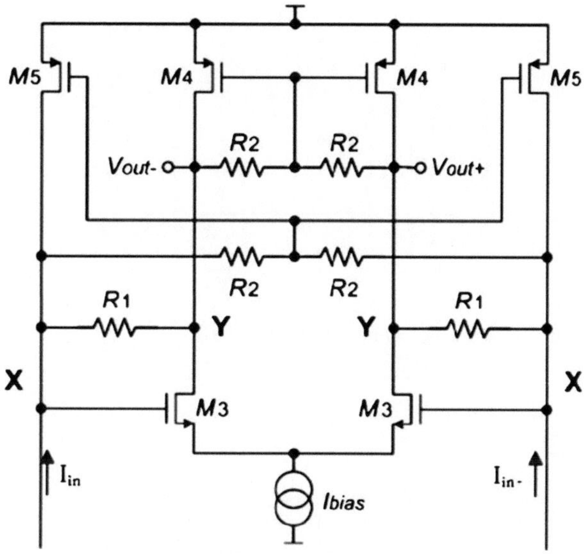

Fig. 3.8 Transimpedance amplifier for the implementation of the summing block [39]

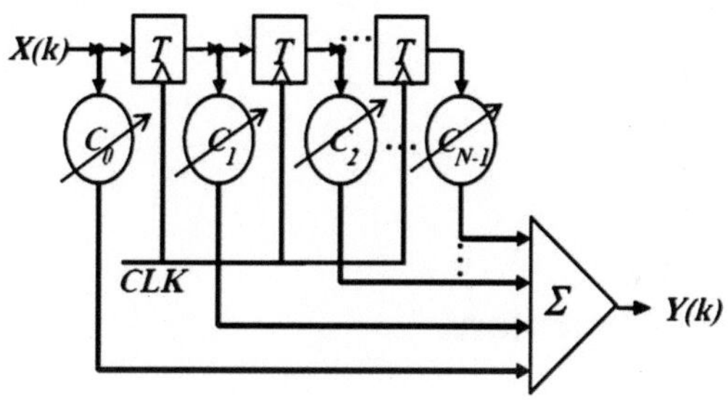

Fig. 3.9 Discrete time FIR filter implementation

multiplied by weighting factors (C_k). The outputs from the weights are summed to produce the transmitted pre-equalized output (y_k) The number of taps depends on the length of the channel relative to the unit interval of the data. The equalizer vector coefficients C(z) are:

$$C(z) = \sum_{k=0}^{N} c_k \, z^{-k} \tag{3.13}$$

The delay units can be implemented by a dynamic True Single-Phase Clocked (TSPC) latch as in Fig. 3.10 [40]. The output combiner circuit, shown in Fig. 3.11, makes the summation and gives the required weights for the delayed signals.

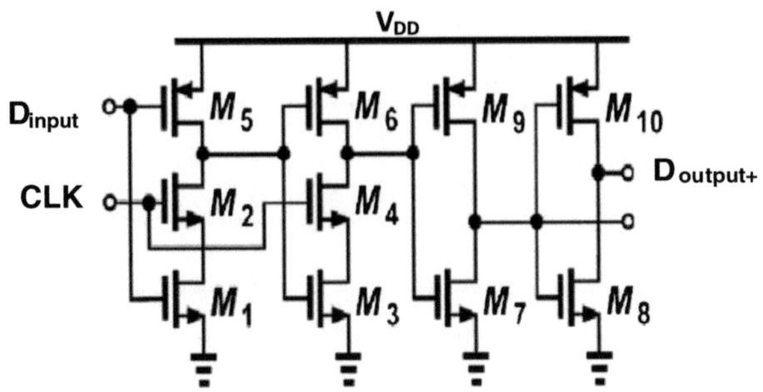

Fig. 3.10 Delay unit cell using a dynamic TSPC latch [40]

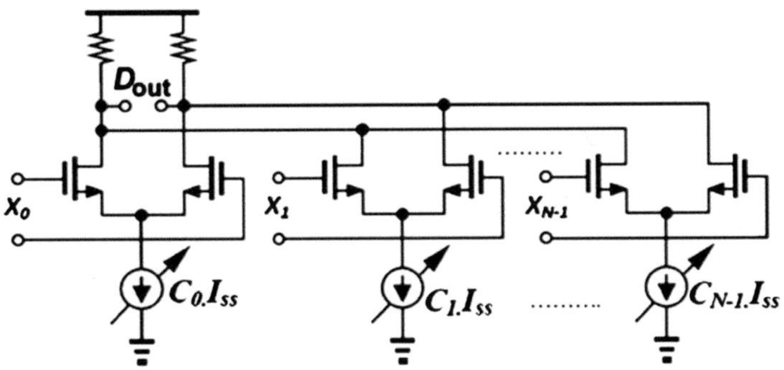

Fig. 3.11 The output combiner

3.5 Equalization Algorithms

There are many factors being important for the choice of an equalization structure and its algorithm like the costs of the computing platform, the power budget, the channel characteristics and the channel data rate. The number of taps used in the equalizer design depends on the maximum expected time delay spread of the channel. The circuit complexity and processing time increases with the number of taps and delay elements.

A substantial increment of the noise power is created using zero forcing equalizers. Least Mean Squares (LMS) and recursive least squares (RLS) algorithms give better performance. Recent techniques for adaptive blind algorithm like Constant Modulus Algorithm (CMA) used for constant envelope modulation and Spectral Coherence Restoral Algorithm (SCORE) exploit spectral redundancy in the transmitted signal to minimize the MSE and to reduce the bit error rate.

1. Zero Forcing Algorithm (ZF)

By using a Zero Forcing (ZF) algorithm the ISI can be completely removed, without taking in consideration the resultanting noise enhancement.

$$C(z) = 1/h(z) \tag{3.14}$$

where h(z) is the channel transfer function in the Z domain. The zero-forcing equalizer removes all ISI, and is ideal when the channel is noiseless. However, when the channel is noisy, the zero-forcing equalizer will amplify the noise greatly. A more balanced linear equalizer in this case is the mean-square error equalizer, which does not usually eliminate ISI completely but instead minimizes the total power of the noise and ISI components in the output. From the point-of-view of minimizing the error probability, it is advantageous to allow some residual ISI if this can reduce the noise power. The Minimum Mean-Squared Error algorithm (MMSE) attempts to minimize the mean squared error (MSE) between the equalizer output and the transmitted data symbol.

$$E\left\{\left(I^\wedge(k) - I(k)\right)^2\right\} = ||e(k)||^2_{min} \tag{3.15}$$

where $e(k)$ is the error at the current sample 'k' and the filter transfer function becomes:

$$C(z) = \frac{1}{h(z) + N_o}. \tag{3.16}$$

2. Least mean square (LMS) algorithms

Least mean square (LMS) algorithms are a class of adaptive algorithms used to optimize a desired filter performance by finding the filter coefficients that relate to producing the least mean squares of the error signal (difference between the desired and the actual signal):

$$E\{|e(k)^2|\} \tag{3.17}$$

where $e(k)$ is the error at the current sample 'k' and $E\{\cdot\}$ denotes the expected value. The LMS algorithm has a simple computational complexity but it has a slow convergence.

3. Recursive least square (RLS) algorithms

The recursive least square (RLS) adaptive algorithm recursively finds the filter coefficients that minimize a weighted linear cost function relating to the input signals. RLS exhibits extremely fast convergence. However, this benefit comes at the cost of high computational complexity.

The cost function we desire to minimize is a weighted least squares error function [42]:

$$Cost(c_n) = \sum_{i=0}^{n} \lambda^{n-i} e^2(i) \tag{3.18}$$

where $1 > \lambda > 0$ is the "forgetting factor" which gives exponentially less weight to older error samples. The smaller λ is, the smaller the contribution of previous samples is. This makes the filter more sensitive to recent samples, which means more fluctuations in the filter coefficients. The RLS convergence is fast and has a good tracking ability but it needs a high computational complexity.

3.6 Adaptive Equalization

If the channel is frequency selective, the equalizer enhances the frequency components with small amplitudes and attenuates the strong frequencies in the received frequency response. For a time-varying channel, an adaptive equalizer is needed to track the channel variations.
Ideally, tap coefficients are tuned to each system to account for operational (V, T) and manufacturing variation. This is done using adaptive algorithms. Perfect adaptation is not practical limited by things like update rate and coefficient resolution.
A filter with N delay elements has N + 1 taps, and N + 1 tunable complex weights. These weights are updated continuously by an adaptive algorithm. The object is to adapt the coefficients to minimize the noise and intersymbol interference (depending on the type of equalizer) at the output. The adaptation of the equalizer is driven by an error signal.

3.6.1 Continuous Time Adaptive Equalizer

Figure 3.12 shows the block diagram for the adaptive analog equalizer introduced in [41]. The received input signal first goes through the equalization filter, which has two signal paths: the low frequency path and the high frequency path. The high frequency path compensates for the high-frequency loss of the channel so that the overall loss becomes equal for all frequencies. The output of the equalization filter is then fed into a comparator. The servo loop determines the need for more boosting by comparing the high-frequency contents of the comparator's input and output signals. Two pairs of high-pass filter and rectifier extract the high-frequency power of these signals, and an error amplifier calculates their difference. After being filtered by a loop capacitor C_2, this difference controls the high-frequency gain, which effectively adjusts the zero frequency of the equalizing filter. Two pairs of low-pass filter and rectifier extract the low-frequency power of the regulating comparator's input and output signals. Then the difference between these two quantities is amplified by the error amplifier. The difference signal is filtered by C_1 and the filtered signal controls the low-frequency gain of the equalizing filter (α).

3.6.2 Discrete Time Adaptive Equalizer

There are two modes that adaptive equalizers work:

- **Training Mode**: To make equalizer suitable in the initial acquisition duration, a training signal is needed. In this mode of operation, the transmitter generates a data symbol sequence known to the receiver.

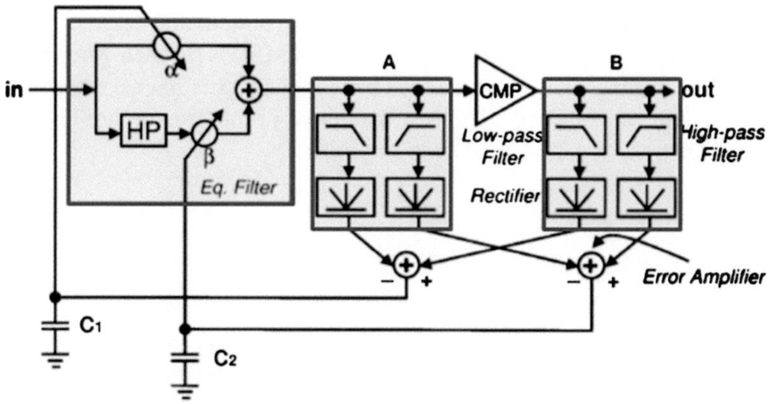

Fig. 3.12 Block diagram for the continuous time adaptive equalizer [41]

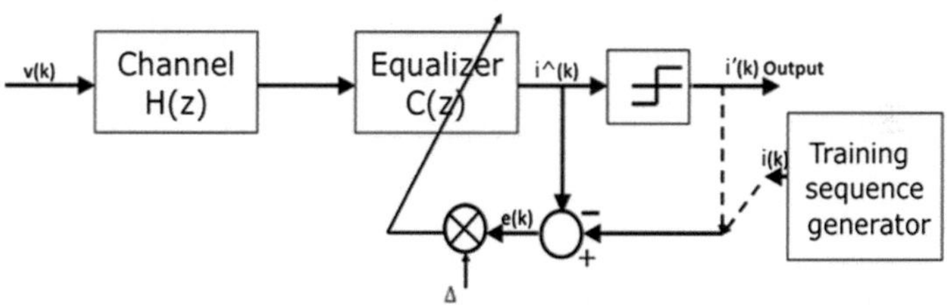

Fig. 3.13 Block diagram of an adaptive linear equalizer

- **Decision Directed Mode**: The receiver decisions are used to generate the error signal. Decision directed equalizer adjustment is effective in tracking slow variations in the channel response. Once an agreed time has elapsed, the slicer output is used as a training signal and the actual data transmission begins.

The forward and feedback filter coefficients are adjusted simultaneously using least mean square (LMS) algorithm to minimize the MSE. The equalizer coefficient vector C in Fig. 3.13 is obtained according to the relation [22]:

$$c_j(k+1) = c_j(k) - \Delta G(k), \quad k = 0, 1, 2, \ldots N \tag{3.19}$$

$$\text{Gradient factor } G(k) = -E(e(k)V^*) \tag{3.20}$$

$$e(k) = I(k) - I^\wedge(k) \text{ (Training mode)} \tag{3.21}$$

$$c_j(k+1) = c_j(k) + \Delta e'(k)v^*(k-j) \tag{3.22}$$

$$e'(k) = I'(k) - I^\wedge(k) \text{ (Decision mode)} \tag{3.23}$$

Δ indicates scale factors that control the rate of adjustment and $e(k) \cdot V^*(k-j)$ is an estimate of the cross-correlation. The minimum MSE is reached when G(k) = 0. Following the training period, after which the equalizer coefficients have converged to their optimum values, the decisions at the output of the detector are reliable to be used to continue the coefficient adaptation process.

3.7 Nonlinear Equalization

The linear equalizer cannot distinguish between the signal and the noise. Both the high frequency signal and the high frequency noise is amplified. The data rate will be limited by signal to noise ratio. Noise becomes a primary design consideration with linear equalizers. An alternative to linear equalization is non-linear equalization. Non-linear equalization is commonly used in applications where the channel distortion is too severe. There are two effective non-linear equalization methods:

- Decision Feedback Equalization (DFE)
- Maximum Likelihood Sequence Estimator (MLSE)

3.7.1 Decision Feedback Equalizers (DFE)

The DFE has a feedback from the filter output to the input, Fig. 3.14. The input signal is summed with the feedback signal to provide input to a decision circuit (slicer), which decodes the analog input signal to a binary signal with a comparator. The output from the slicer is used as input to the FIR filter. The basic idea is that if the values of the symbols already detected are known (past decisions are assumed correct), then the ISI contributed by these symbols can be canceled exactly. The error in current decisions will corrupt future decisions. There are coding methods to minimize this impact. The DFE corrects only for post-cursor ISI. The DFE requires an input signal with open eye, so it is typically used with a feed forward linear equalizer on the front end of the DFE.

Adaptive DFE Equalization

The forward and feedback filter coffiecients for the adaptive DFE (Fig. 3.15) are adjusted simultaneously to minimize the MSE using the steepest decent algorithm [22]:

$$c(k+1) = c(k) + \Delta E(e(k)V^*(k)) \tag{3.24}$$

$$e(k) = I(k) - I^\wedge(k) \text{ (Trainingmode)} \tag{3.25}$$

$$c(k+1) = c(k) + \Delta e'(k)V^*(k) \tag{3.26}$$

$$e'(k) = I'(k) - I^\wedge(k) \text{ (Decisionmode)} \tag{3.27}$$

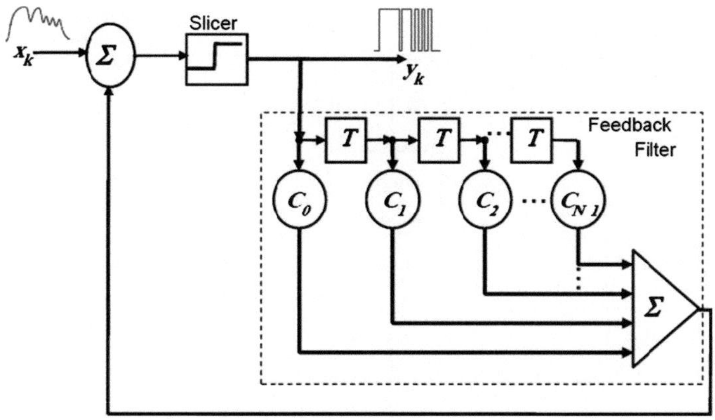

Fig. 3.14 Decision feedback equalizer block diagram

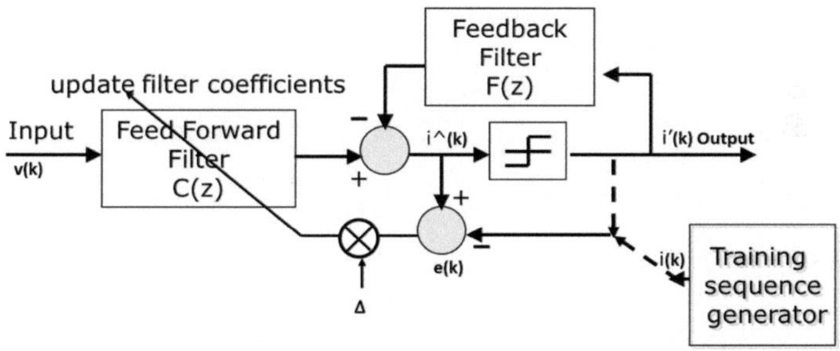

Fig. 3.15 Adaptive decision feedback equalizer block diagram

3.7.2 MLSE Nonlinear Equalization

In fact, it turns out that all the equalization methods discussed in the previous section are not optimal. Because of the fact that the effect of a symbol is spread to other symbols, it is intuitive that the optimal receiver should observe not only the segment of received signal concerning the desired symbol, but the whole received signal instead. Instead of deciding a transmitted symbol at a time, we can consider to decide the whole transmitted symbol sequence simultaneously from the received signal. In this way, we aim at minimizing the probability of choosing the wrong sequence of symbols instead of the average symbol error probability. With this sequence detection approach, we can employ the Maximum Likelihood principle to achieve our goal. The resulting "optimal" receiver is referred to as the Maximum Likelihood Sequence Estimator (MLSE) receiver. MLSE tests all possible data sequences (rather than decoding each received symbol by itself), and chooses the data sequence with the maximum probability as the output. The block diagram of an MLSE receiver is shown in Fig. 3.16. The MLSE receiver can be implemented with the Viterbi algorithm [43, 44]. The Viterbi algorithm can be simply described as an algorithm which finds

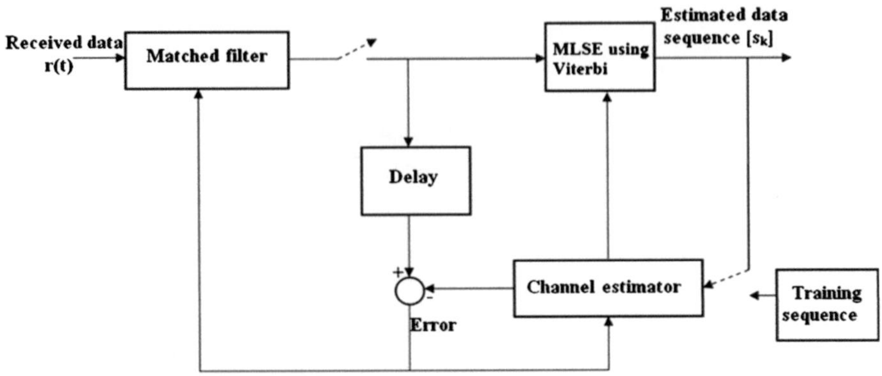

Fig. 3.16 Maximum likelihood sequence estimator (MLSE) block diagram

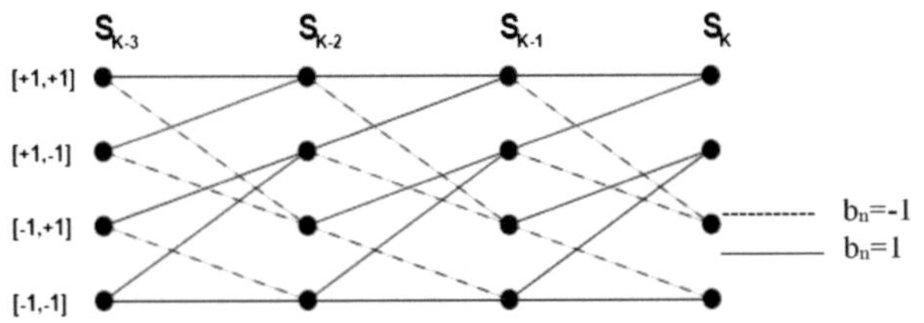

Fig. 3.17 Trellis diagram for L = 2

the most likely path through a trellis, i.e. shortest path, given a set of observations. The trellis in this case represents a graph of a finite set of states from a Finite States Machine (FSM). Each node in this graph represents a state and each edge a possible transition between two states at consecutive discrete time intervals. An example of a trellis is shown below in Fig. 3.17.

MLSE usually has a large computational requirement. MLSE requires knowledge of the channel characteristics in order to compute the metrics for making decisions. MLSE also requires knowledge of the statistical distribution of the noise corrupting the signal. The MLSE for a channel with ISI has a computational complexity that grows exponentially with the length of the channel time dispersion. If the size of the symbol alphabet is M and the number of interfering symbols contributing to ISI is L, The Viterbi algorithm computes M^{L+1} metrics for each received symbol. In most channels of practical interest, such a large computational complexity is expensive to implement [22]. For time-invariant channel impulse response, the received symbols are described as

$$\mathrm{r(k)} = \sum_{j=0}^{L-1} h_j \, b_{k-j} + n_k \tag{3.28}$$

where r_k is the received symbol, b_k denotes the actual transmitted sequence, n_k is the Gaussian noise, and h_j is the coefficient of the estimated channel impulse response. The equalizer is responsible for reversing the effect of the channel on the transmitted symbols in order to produce the sequence of transmitted symbols with maximum confidence. To optimally estimate the transmitted sequence of length M in an optical communication system, the least mean square error function must be minimized.

$$\sum_{k=1}^{M} \left| \mathrm{r(k)} - \sum_{j=0}^{L-1} h_j s_{k-j} \right|^2 \tag{3.29}$$

Here, $\mathbf{s} = [s_1, \ldots, s_M]$ is the most likely estimated transmitted sequence that will maximize the probability of $P(\mathbf{s} \mid \mathbf{r})$. The Viterbi MLSE equalizer is able to solve this problem exactly, with computational complexity linear in M and exponential in L [20].

The update relations of the states are depicted by the trellis diagram. The trellis diagram for the case of $L = 2$ is given in Fig. 3.17. In a trellis, each node corresponds to a distinct state at a given time(S), and each edge represents a transition to some new state at the next instant of time. The trellis begins and ends at the known states S_0 and S_k. Its most important property is that to every possible state sequence S there is a unique path through the trellis, and vice versa.

3.8 Performance Comparison of Equalization Methods

For purposes of comparison, the BER is calculated for different equalization methods using a channel that introduces severe ISI. The impulse response for a 50 m SI-POF channel is (0.18, 0.43, 0.27, 0.1, 0.02) and the sampling frequency is 312.5 MHz [45]. The normalized frequency response for 50 m SI-POF channel is shown in Fig. 3.18. The adaptive linear equalizer uses a 7 taps and the DFE uses three feed forward and three feedback taps. The RLS algorithm is used to initially set the weights, and the LMS algorithm is used thereafter for speed purposes. The RLS algorithm is used for the first block of data to ensure rapid tap convergence. The LMS algorithm is used thereafter to ensure rapid execution speed.

The MLSE equalizer trace back length is six. The assumed estimated length of the channel impulse response is one sample longer than the actual length. The code for all above equalization methods built in MATLAB and the BER is calculated for different signal-to-noise ratios.

The BER for the equalized channel versus signal-to-noise ratio for different equalization techniques for 50 m POF channel are plotted in Fig. 3.19.

For the linear equalizer note that as the SNR increases, the linearly equalized signal has a lower BER. But the slop of the BER decreases with a small rate; this highlights the fact that a linear equalizer must have many more taps to adequately equalize the POF channel.

Note that the DFE is much better able to mitigate the POF channel than the linear equalizer, as shown in the BER plot. But the error is still larger than that with the

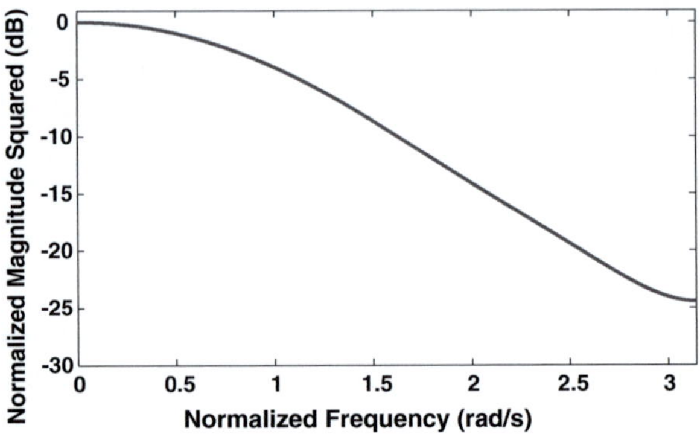

Fig. 3.18 Normalized frequency response of a 50 m SI-POF channel

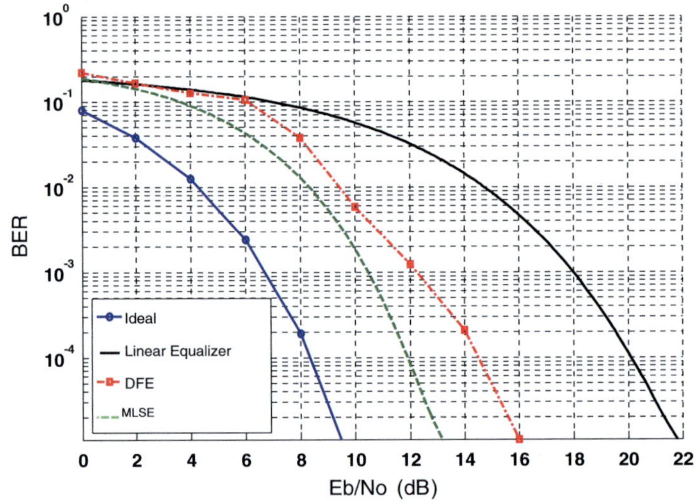

Fig. 3.19 BER versus signal-to-noise ratio of different equalization techniques for a 50 m POF channel

MLSE equalizer due to the DFE error propagation caused by feeding back detected bits instead of correct bits.

The MLSE equalizer has the lowest BER compared to the linear and the DFE equalizer. The MLSE equalizer has the highest complexity and needs a larger processing time compared to the linear and the DFE equalizer.

For BER $= 10^{-4}$, the MLSE shows a 3 dB improvement in the signal to noise ratio over the DFE whereas the DFE shows a 5 dB enhancement versus the linear equalizer.

A low complexity implementation with a good performance for a channel with low levels of ISI is obtained using linear equalizers. In case of channels with severe ISI, DFE and MLSE is the best option. The MLSE equalizer is not appropriate for online processing in real-time.

Chapter 4
High-Speed Transmission over Step-Index PMMA Plastic Optical Fibers

4.1 Why Large-Core PMMA POF?

Polymer Optical Fibers based on Poly-Methyl-Methacrylate with step index 1 mm core diameter have gained interest in the recent years for their interesting properties compared to glass optical fibers (GOF). The main advantages of POF when compared to GOF are:

- POF large core diameter (1 mm) allows do-it-yourself installation and termination with common cutter and electrician-like low-cost tools; besides PMMA material is a very inexpensive material.
- Because POF has a larger diameter than GOF; their connectors are less complex, cost less, and are less likely to suffer damage than connectors for glass optical fibers.
- POF high diameter and numerical aperture make bending loss sensitivity much lower than silica fiber (GOF). Also the high numerical aperture makes the coupling loss from the light to POF smaller than that of GOF.
- POF mechanical resilience and elasticity makes it possible to step on it and even tie it.
- Dust and water harm POF to a much smaller extent than GOF.
- The optical sources for POF are in the visible range, and the optical launch is usually non collimated. POF optical sources are thus intrinsically eye-safe, as the signal can be seen by the naked eye.

For all these reasons, POF is potentially very interesting in several applications (industrial automation, automotive, home networking) where it shows key advantages to the more traditional copper cabling:

- Complete immunity to electromagnetic interference (EMI)
- Being an electrical insulator (like GOF), POF can be laid down in power ducts.
- Lower weight (a fundamental issue in the automotive sector).
- POF transceivers require less power than copper transceivers.

M. Atef and H. Zimmermann, *Optical Communication over Plastic Optical Fibers*, Springer Series in Optical Sciences 172, DOI: 10.1007/978-3-642-30388-3_4,

Table 4.1 The requirements for 100 Mbit/s and 1 Gbit/s PMMA POF systems [1]

Feature	100 Mbit/s system	1 Gbit/s system
1 mm PMMA SI-POF length	50 cm–100 m	50 cm–100 m
Maximum DR	125 Mb/s	1.25 Gb/s
BER	$<10^{-12}$	$<10^{-6}$
Temperature range	0 to +60 °C	0 to +60 °C
Eye safety	Class 1	Class 1
Maximum power consumption	<0.4 W per port transceiver <3.5 W in full operation media converter	<3 W per port in full operation media converter
Standby mode	<0.5 W	<0.5 W

These native properties have to be balanced by some drawbacks:

- The POF modal dispersion limits the POF bandwidth to 40 MHz for 100 m length
- PMMA exhibits a strong attenuation, minimum for visible light (150 dB/km at 650 nm, to be compared to 0.25 dB/km at 1550 nm for silica single-mode fiber). This limits the reach of the links to short reach and low DR transmission [1, 2]
- Lack of high-temperature POF fibers (125 °C).

4.2 POF Main Applications

Today on the market several suppliers offer PMMA POF media converter solutions at 100 Mbit/s. With such performance PMMA fiber may be used in the home to interconnect all devices usually communicating through Fast Ethernet interfaces; for example the link between the home gateway and the Set Top Boxes (STB). The requirements for 100 Mbit/s and 1 Gbit/s PMMA POF systems are illustrated in [1] (Table 4.1).

The two major applications of POFs are in the industrial control and automotive fields. Until last year industrial controls remained the biggest and most stable market for the POF industry. The main driver for POF in the industrial-control market is the need for data links that resist EMI. The POF sales to automotive companies rose to become the single largest source of revenue for POF makers. Today, the major source for the POF business lies in the use of POF products by automobile companies. Media Oriented Systems Transport (MOST) is a vehicle bus standard, intended for interconnecting multimedia components in automobiles and other vehicles. The original MOST system was designed for 25 Mb/s and increased to 50 Mb/s. The next generation is planned to increase to 150 Mb/s [2].

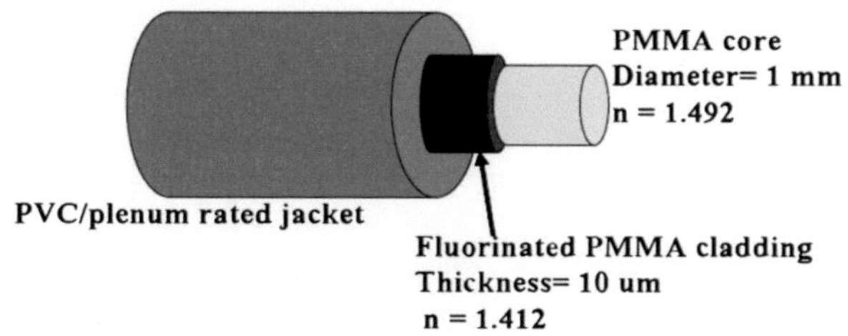

Fig. 4.1 The structure of standard PMMA SI-POF

4.3 Characteristics of Step-Index PMMA POF

The 1 mm diameter PMMA core is enclosed by a thin cladding (10 μm) out of a fluorinated polymer. The PMMA core has an index of refraction of 1.492 and the cladding has a refraction index of 1.412. Through total reflection the light is guided inside the core, Fig. 4.1. The lowest attenuation values occur at 520 nm (green), 560 nm (yellow) and 650 nm (red), Fig. 4.2. The difference in the transit time of different rays in the core leads to a broadening of the light pulses launched. Possible bit rates are limited to about 100 Mbit/s over 100 m for SI types. Modified index profiles allow higher transmission data rates. The numerical aperture (NA) for the standard PMMA SI-POF is 0.5. The maximum operation temperature ranges from +70 up to +85 °C depending on the application.

4.3.1 POF Frequency Response and Bandwidth

A pulse, during its propagation through the fiber, experiences a spread known as dispersion.

Modal dispersion affects multimode fibers like POF. It is caused by the different modes or paths followed by light rays in the fiber, which gives rise to different ray transit times. Therefore, in any communication link through a POF it is observed that the dispersion is the parameter that determines the maximum bandwidth that can be transmitted through the optical fiber.

The POF bandwidth depends on the modal dispersion and its evolution when passing through the fiber. Higher order modes are more attenuated, and mode mixing effects mainly from imperfections at the core cladding interface lead to the evolution of a so-called equilibrium modal distribution (EMD) after a certain fiber length. The equilibrium is reached for distances around 15 m when the NA is large. If the NA is small, then the equilibrium is reached much later [46, 47].

Experimental measurements and mathematical models for the frequency responses for different PMMA SI-POF lengths and types are discussed in [48, 49]. Also a detailed study for effect of POF length on the bandwidth can be found there. The following POF types are used in the measurements: ESKA-PREMIER GH4001 (GH)

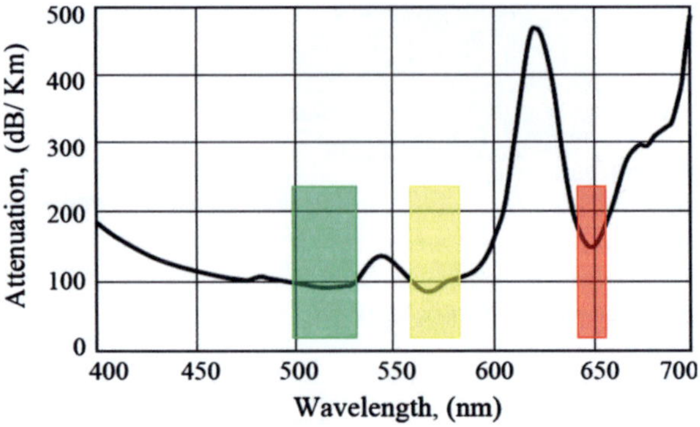

Fig. 4.2 Typical loss curve for a PMMA POF

from Mitsubishi, HFBR-RUS100 (HFB) from Agilent, and PGU-FB1000 (PGU) from Toray. The PMMA SI POF bandwidth is nearly 40 MHz for 100 m length which limits the DR to less than 100 Mb/s over 100 m by using the binary modulation.

We measure the frequency response for 1 mm core diameter PMMA SI-POF with NA = 0.5 (POF1.100.B22) at different POF lengths 20, 40, 115, 200 m. The network analyzer (HP 8753E) was used to modulate 655 nm LD. The POF was butt coupled to the LD without the use of any lens. The optical power was received by a 12 GHz photoreceiver (New Focus Model 1554). The measured SI-POF frequency responses for different POF lengths are plotted in Fig. 4.3. The changing of the POF length has two effects on the frequency response. Firstly, the bandwidth is decreased by increasing the POF length.

Secondly, the roll-off is also increased by increasing the POF length. The measured electrical bandwidth, optical bandwidth and roll-off are illustrated in Table 4.2.

For POF lengths less than 20 m the frequency roll-off is smaller than 4.2 dB/octave, using binary signal is better than using 4-PAM signal for these lengths because the 4-PAM has a power penalty of 5.1 dB (relation (2.15)) which is higher than the POF roll-off by 0.9 dB. For POF lengths larger than 40 m the roll-off is larger than 6 dB/octave, 4-PAM signal is better than binary signal for these lengths because the 4-PAM has a power penalty of 5.1 dB which is larger than the roll-off by 0.9 dB. For POF lengths larger than 40 m and less than 115 m the roll-off is less than 7.2 dB/octave, 4-PAM modulation is better than binary for these lengths because the 4-PAM has a net 2.1 dB improvement compared to binary. For POF lengths larger than 115 m and less than 200 m the roll-off is less than 8.4 dB/octave, 4-PAM signaling beats binary for these lengths by 3.3 dB improvement compared to binary. As a conclusion if the system bandwidth is limited by the POF; it is preferred to use binary for POF lengths less than 40 m to save transmitted optical power for lengths longer than 40 m it is preferred to use 4-PAM signaling to save optical power.

The Shannon capacity, Eq. (2.2), can be calculated for different POF lengths by using the measured electrical bandwidth for each POF length in Fig. 4.4. The laser source has 655 nm wavelength and 6 dBm optical power. The used POF has 0.15 dB/m

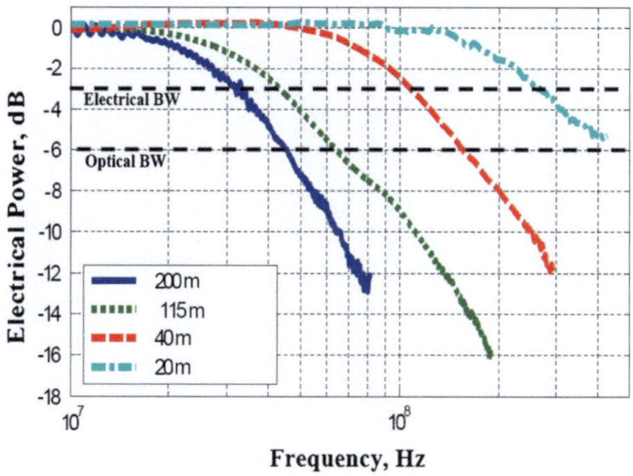

Fig. 4.3 Measured standard PMMS SI-POF frequency response

Table 4.2 Standard PMMA SI-POF measured electrical bandwidth, optical bandwidth and roll-off

	20 m	40 m	115 m	200 m
Electrical BW (MHz)	256	108	43	31
Optical BW (MHz)	500	156	64	45
Roll-off (dB/decade)	14	20	24	28
Roll-off (dB/octave)	4.2	6	7.2	8.4

attenuation at 650 nm wavelength. The optical receiver has 622 MHz bandwidth and −35 dBm input noise power. For low data rate a 112 MHz receiver was used with a −42 dBm input noise power. The calculated Shannon capacity as a function of POF length is illustrated in Fig. 4.4. The region 4 cannot be reached due to the limited received optical power and POF bandwidth. It can be reached by increasing the optical power and using equalization.

Due to the roll-off of the system (POF and optical receiver) the binary signaling is the best format in region 1, 4-PAM is better than binary in region 2 and 8-PAM is preferred in region 3.

4.3.2 POF Bending Loss

Macro bends are mainly produced during the POF installation process. When the curvature is small, losses are practically negligible. As the bend radius decreases radiation losses increase, but the losses being higher for GOFs than for SI-POFs of the same diameter. For example, the total radiation loss caused by a full turn in a typical PMMA SI-POF of 1 mm diameter and NA = 0.5 for bend radii 5 and 15 mm

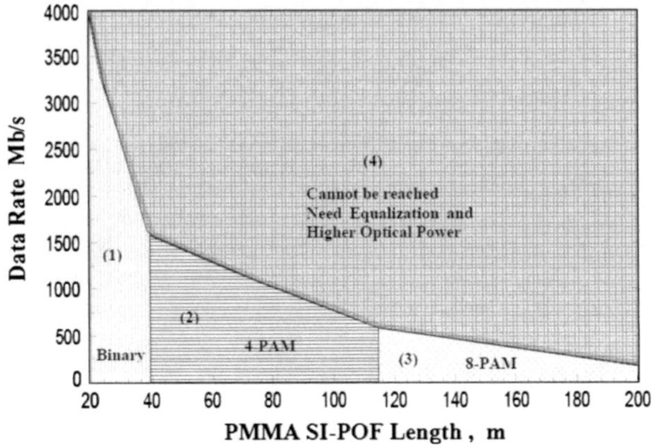

Fig. 4.4 Shannon capacity as a function of POF length

are 2.1 and 0.5 dB respectively. The power decay when reducing the bend radius is exponential, with an exponent that depends on various parameters such as bending radius, light wavelength and NA [48, 49].

The NA is the Sine of the acceptance angle δ_A. The acceptance angle defines the maximum angle where light it guided through the fiber. It can be calculated with the refractive indices of the core and the cladding with the following formula (4.1).

$$NA = \eta_o \, sin \, \sigma_A = \sqrt{\eta_{core}^2 - \eta_{cladding}^2} \tag{4.1}$$

where n_0 is the index of refraction of the medium in which the light is entering (1.0 for air). A high NA (corresponding to high possible propagation angles) makes the light launch into the fiber easier and reduces the additional loss at fiber bends, on the other hand, increases the mode dispersion too. Since the bandwidth is proportional to the fiber NA, the transmission capacity can be increased by lowering the NA. On the other hand, lowering NA causes increased bending sensitivity.

4.3.3 POF Coupling Loss

The typical numerical aperture of silica fibers are NA = 0.2 for step-index and graded-index types and NA = 0.1 for single mode types. The numerical aperture of 1 mm POFs can reach to 0.5 and the acceptance angle is about 30°.

A high NA (0.5 for 1 mm PMMA SI-POF) makes the light launch into the fiber easier and allows simple coupling from the light source to the POF. The guaranteed output power from VCSEL and LD is much higher than that of RC-LED and worst for LED. The reason for this is a more efficient coupling from VCSEL or LD to POF. This advantage was dropped with the LED for reasons of cost. An adjusted LED

Table 4.3 Different methods for coupling 1 mm diameter POF to photodiode [49]

PD diameter (μm)	Butt coupling (%)	Lens (%)	Dielectric taper (%)
300	8.3	31.6	50.5
400	16.4	46.6	75.5
500	23.4	60.6	83.2

or RC-LED with optimized coupling e.g. with micro-mirror, could guarantee a high fiber coupling optical power.

Coupling the optical power from 1 mm POF to a PD is not an easy task. The output optical power from the POF has an angle within the range of ±30. This means that a large area PD will be needed to be relatively close to the POF in order to capture all of the light. One main problem is that large-area PD means a large photodiode capacitance which limits the bandwidth of the optical receiver. The bandwidth can be increased by decreasing the transimpedance amplifier feedback resistance to obtain the required bandwidth. Decreasing the feedback resistance has the effect of increasing the noise and decreasing the optical receiver sensitivity. So, a larger PD area has the bad effect of limiting the bandwidth or the sensitivity of the optical receiver. This leads to use a small-area PD to which the POF is coupled via a lens or taper. A comparison between the coupling efficiencies for lens and taper for different PD diameters shows that the coupled power can be increased by using lens or taper; see Table 4.3 [49]. The taper has a better coupling efficiency than the lens and it is promising for coupling 1 mm POF to smaller diameter PDs.

4.4 Light Sources for PMMA POF

LEDs are large-area emitting components which are available at all wavelengths (520, 570, 650 nm are available for PMMA POF). Due to their large area they cannot be modulated very fast (250 Mb/s). Due to their wide emission angle (50°) the coupling loss to POF is high. LEDs are low cost and simple to use. Resonant cavity LED is different from conventional LED (RC-LED due to the two additional semiconductors Bragg mirrors above and below the light generation layer. Nowadays, RC-LEDs are available in the 650 nm wavelength range. Due to the RC-LEDs special design they allow higher modulation speed (500 Mb/s) than LED. Conventional laser diodes (LD) emit collimated light from the edge with a small emission angle which enhances the coupling efficiency to the fiber. They have available wavelengths higher than 635 nm (650 nm is available). A green laser diode with wavelength 520 nm will be commercially available in the next years; engineering samples are available now from OSRAM. A blue-green laser with a wavelength of 488 nm became recently available from Nichia.

Laser diodes are very fast (2 Gb/s), but they operate only if there is a threshold current, which is changing strongly with the temperature. For the operation in data communication an automatic power control (APC) is required, which makes the

Table 4.4 Light sources for PMMA POF

Parameter	LED	LD	VCSEL	RC-LED
Wavelength (nm)	520, 570, 650	650, 520, 488	660	650
Threshold current	No	>15 mA	2 mA	No
Optical power (mW)	2	up to 20	1	2
Data rate	250 Mb/s	2 Gb/s	5 Gb/s	500 Mb/s
Optical power sensitivity to temperature (dB/°C)	−0.01, −0.05, −0.01	−0.02	−0.08	−0.03
Wavelength sensitivity to temperature (nm/°C)	0.04, 0.12, 0.12	0.18	0.06	0.07
Spectral bandwidth (nm)	40, 25, 30	2	1	10
Cost	Very cheap	650 nm is cheap; the others are expensive	Expensive	Cheap
Operating temperature (°C)	85	85	50	85

components more difficult. VCSELs radiate like LEDs from the upper surface. But the light is collimated almost like an edge emitting laser diode which leads to a better coupling efficiency and smaller lunching NA to the POF. There is a threshold current, but a very small one (typically a few milliamps). VCSEL is the fastest light source which can work up to DR 5 Gb/s [50]. The important characteristics for LED, RC-LED, LD and VCSEL are illustrated in Table 4.4.

4.4.1 Light Source Modulation

In optical data communication there exist two main principles to implement a light source which contains the data signal as amplitude information. The first is a CW with external modulator and the second method is a direct modulated light source.

The CW light source with the external modulator is commonly used in long range optical data communication in the 1500 or 1300 nm wavelength range. The advantage of this modulation technique is that the spectral purity of the emitted modulated light is optimal high, since the laser source works in a constant bias point. The spectral purity is especially in long range applications an important topic, since chromatic dispersion limits the useful bandwidth and so the data rate of the link.

Direct modulated light sources offer a simple and low-cost solution for short range data communication, where chromatic dispersion is not a dominant problem. Additionally fast external modulators are not yet available for the near infrared (IR) up to the ultraviolet (UV) wavelength range and far too expensive for low cost applications. Possible direct switchable light sources are LEDs for data rates up to 250 Mbps and RC-LEDs for data rates up to 500 Mbps. For higher data rates laser

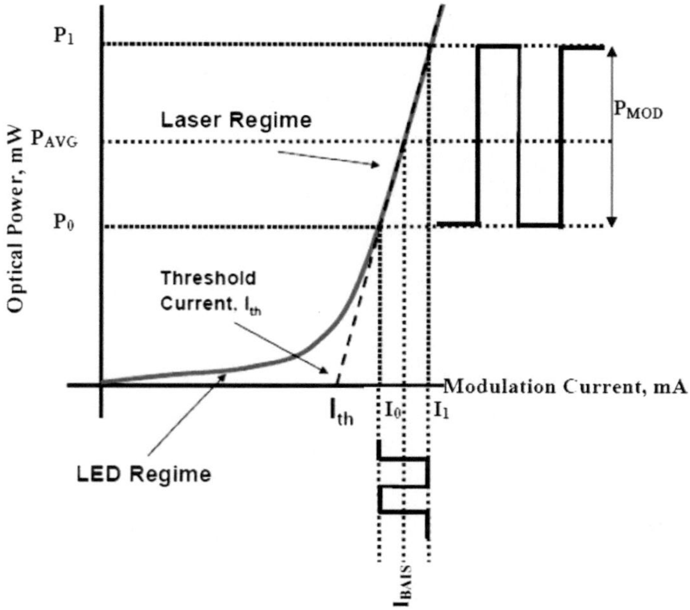

Fig. 4.5 Principle of laser diode modulation with electrical current

diodes are the preferred sources. There exist two main types of laser diodes: edge emitting laser diodes and vertical surface emitting diodes.

4.4.2 Laser Diode Modulation

The laser diode power is modulated by changing the current through the laser diode, the principle is shown in Fig. 4.5.

The conversion factor between modulation current and modulation output power in the laser operating mode is called slope efficiency, S_{laser} (in units of W/A). The slop efficiency is the ratio of increase in light power intensity ΔP to the increase in injection (modulation) current ΔI, which is given by [51]:

$$S_{laser} = \Delta P/\Delta I \tag{4.2}$$

The minimum optical power is chosen to be larger than zero to improve signal response of the laser diode, regarding response time and overshoot. The ratio between maximum optical power and minimum optical power is called extinction ratio (ER) [52, 53] and can practically range from 2 to 10.

$$ER = \frac{power(1)}{power(0)} \tag{4.3}$$

Another important parameter of a laser source is the average emitted laser power P_{AVG} [53].

$$P_{AVG} = \frac{power(1) + power(0)}{2} \tag{4.4}$$

The relation (4.5) calculates from extinction ratio and average optical power the modulation of the power. This is useful since the extinction ratio is usually known from the source and only average optical power can easily be measured [53].

$$P_{MOD} = 2P_{AVG}\frac{ER - 1}{ER + 1} \tag{4.5}$$

In the ideal case the extinction ratio would be infinite. In this case, the maximum modulation is reached for a given average power so relation (4.5) is reduced to $P_{MOD}(ER \to \infty) = 2P_{AVG}$. Due to the finite extinction ratio, the laser source has a power penalty compared to the ideal case with the same modulation swing. The power penalty in dB can be calculated from relation (4.6), and gives a power penalty of 1.76 dB for an extinction ratio of 5 [53].

$$\text{ER power penalty} = 10Log\frac{ER + 1}{ER - 1} \tag{4.6}$$

4.4.3 Laser Nonlinearity

In optical communication systems employing directly modulated laser LD there is a distortion due to the LD nonlinearity in its input current versus output power curve. This nonlinearity has no effect on the performance when binary modulation is used, so nonlinearity compensation is not needed. A different situation appears when using systems based on multilevel modulations, such the 4-PAM that we needed to adopt in our system. As an example, Fig. 4.6 presents the eye diagram resulting at the output of the red LD used in our system, when driving its input with an ideal 4-PAM signal. In this curve, the baud rate was limited to 125 Mbaud just to avoid mixing LD nonlinear effect and bandwidth limitations effect. This result shows that some form of nonlinearity compensation is strictly needed. We compensated for this nonlinearity in the transmitter part (12 bits AWG PCI card), which was anyway able to compensate different LDs in different operating conditions. The compensated signal is illustrated in Fig. 4.7.

4.5 Minimizing the Effects of POF Modal Dispersion

Several simple methods can be used to minimize the effects of modal dispersion. Firstly, by placing the PD at a distance L from the POF fiber end facet, only the lower order modes will be coupled to the PD. The higher order modes are spatially

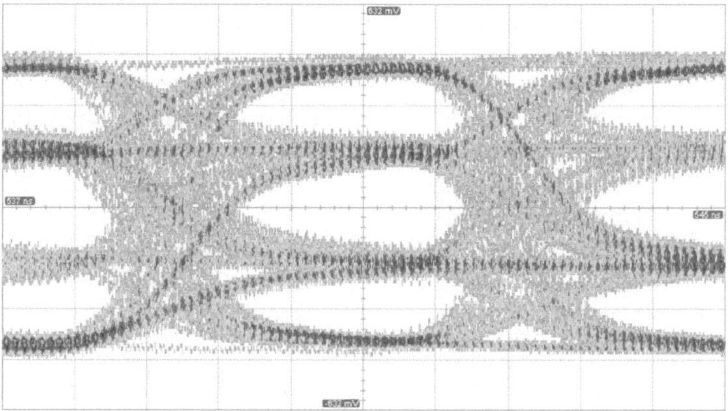

Fig. 4.6 Effect of laser diode nonlinearity of the 4-PAM signal

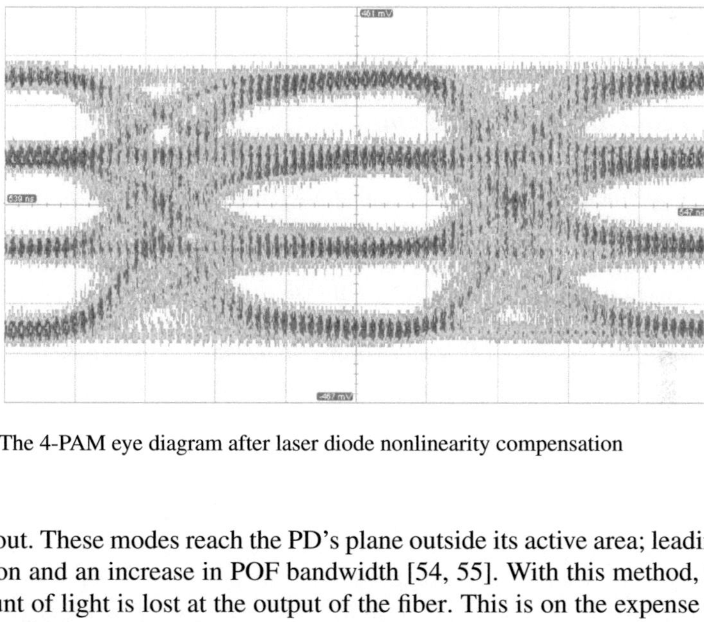

Fig. 4.7 The 4-PAM eye diagram after laser diode nonlinearity compensation

filtered out. These modes reach the PD's plane outside its active area; leading to less dispersion and an increase in POF bandwidth [54, 55]. With this method, however, an amount of light is lost at the output of the fiber. This is on the expense of power budget; a high power laser is needed to compensate this optical power loss. This will make a problem with the eye safety regulations.

A second method to decrease the effect of model dispersion is by using a collimated lens with low NA. A collimated lens is used to collimate the laser light beam with low NA into the SI-POF. This way, only a few low order modes will be excited and modal dispersion will be limited. This requires an extra lens and a precise adjustment to obtain the small lunch angel laser beam.

Another method is to use a pre-equalizer (high pass filter) to peak the laser diode or and post-equalization to compensate for the POF limited bandwidth. Pre-equalizing of light source lowers the light source modulation depth; this reduced the actual power per pulse compared with rectangular pulses without peaking. This is at the expense of system power budget. It follows that the use of pre- or post-equalization methods are of particular interest in systems that have adequate power reserves. Also for fixed pre-equalizer, if the frequency response changes, as a result of different lengths of the POF or a bend in the fiber, the result will be too much or too little compensation so that the BER will increase.

The adaptive equalization is required to have sufficient channel bandwidth at different channel conditions. A combination of a fixed pre-equalizer with a small peaking and an adaptive post-equalizer even is a better solution. The small peaking will have a small effect on the needed transmitted optical power per pulse compared with rectangular pulses without peaking. The fixed pre-equalizer will ease the design of the post-equalizer and help the adaptive post-equalizer to reach the required bandwidth.

4.6 State of the Art for Transmission over Standard PMMA SI-POF

4.6.1 Without Using Equalization

Real-time implementation of a DMT transmitter with an FPGA was demonstrated for low-cost, PMMA SI-POF applications based on commercial RC-LED and large-diameter (540 μm) photodiode [56]. The 1.125 Gbit/s DMT sequence generated by the DAC is used to drive an RC-LED for transmission over 10 m of SI-POF. Such distances are typical for automotive networks and the main limitation originates from the low bandwidth of the RC-LED based transmitter. The received optical power after 10 m SI-POF was −3 dBm and the modulation index was approximately 0.6. A large-diameter (540 μm) photodiode with separate transimpedance amplifier was used to receive the optical signal and a digital storage oscilloscope sampling at 2.5 GS/s is used for demodulation and evaluation of the received DMT sequence.

Simultaneous transmission at a bit-rate of 500 Mb/s per channel with BER $<10^{-10}$ was shown, resulting in a total bit-rate of 1 Gbit/s over 25 m of standard PMMA SI-POF [57]. The used modulation format was Angular Mode Group Diversity Multiplexing (AMGDM) which is based on the principle that the axial angle of light propagating in a SI-POF is preserved for relatively short distances. By launching light from two red laser diodes (658 and 635 nm), each at a different angle, into a standard SI-POF, two different angular mode groups were excited, thereby creating two independent transmission channels in a POF. Moreover, these individual channels will experience less modal dispersion because fewer modes are present in each channel, leading to higher bandwidths. Using a demultiplexing setup consisting of two lenses and a concentric mirror, these two angular channels could be separated and detected properly at the receiver side with two photodiodes with an active diameter of 1 mm.

1.25 Gb/s error-free transmissions over 25 m of standard PMMA SI-POF by special means to improve the bandwidth was reported in [54]. A 1.25 Gbit/s, $2^7 - 1$, NRZ PRBS bit stream was generated and used to drive the laser diode (Hitachi HL6501MG, 658 nm, 30 mW). From the collimating lens, a collimated beam with NA $= 0.11$ and 0.5 mm width is launched into 25 m of SI-POF. At the output of the 1 mm SI-POF, a Si PIN photodiode (Hamamatsu S9055) with a diameter of 200 μm and 1.5 GHz bandwidth was placed to detect the transmitted optical signal. The photodiode was placed at 3 mm distance from the output fiber end to spatially filter out the higher order modes and increase the bandwidth (17.4 dB coupling loss from the POF to the PD due to this method). Via a 900 MHz transimpedance amplifier and a 1000 MHz low-pass filter, the detected signal is passed to the BER test set and also displayed on a sampling oscilloscope. A BER$<10^{-11}$ was reported with a received optical power of -17.4 dBm after 25 m of 1 mm SI-PMMA-POF at a bit-rate of 1.25 Gbit/s.

1 Gbit/s transmission using RC-LED, and 1.5 Gbit/s using VCSELs over 50 m PMMA SI-POF was demonstrated using an optimized DMT modulation with offline processing [59]. Using Matlab a DMT signal was computed and loaded onto an arbitrary waveform generator on a PCI card. The signal is then sent via a biasing circuit to either a VCSEL or a RC-LED emitting both at 650 nm. The optical DMT signal was then transported over 50 m of SI-POF and detected by a high-speed large-area (700 μm diameter) PIN photodiode that includes a Gbit/s bandwidth transimpedance amplifier. The electrical signal is captured by a real time digital sampling oscilloscope. The captured data were then loaded onto a PC and then post-processed with MATLAB. The results stored on a PC were then passed through the Levin-Campello Algorithm (LCA) to determine the achievable data rate and optimum modulation conditions. The corresponding signal is then created on the PC and loaded again on the AWG. The data rates achieved are 1.5 Gbit/s and 948 Mbps, respectively, for the VCSEL and the RC-LED with average BER values of $1.2 \cdot 10^{-5}$ and $1.3 \cdot 10^{-5}$.

1 Gbit/s transmission using a RC-LED, and 1.5 Gbit/s transmissions using a VCSEL over 50 m PMMA SI-POF was demonstrated using an optimized DMT modulation with offline processing [59]. Using Matlab a DMT signal was computed and loaded onto an arbitrary waveform generator (ARB) on a PCI card.

Transmission of 1 Gbit/s over 100 m of SI-POF is demonstrated in [58]. The bandwidth limitation of the SI-POF was compensated by spectrally efficient adaptive multiple subcarrier modulation. The signal was generated by a signal generator (VSG). The complex waveforms have been externally calculated and were repetitively output by the AWG. The generated multiple subcarrier signals were used to modulate a lensed 650 nm LD. A DC-bias was introduced in order to allow for linear transmission. The output light was efficiently coupled into 100 m of SI-POF. The receiver comprised a Si-PIN photodiode with an active diameter of 1 mm followed by a TIA. By using adaptive multiple subcarrier modulation the transmitted signal was optimized for this specific channel characteristic. In the experiment each subchannel was demodulated by using a commercial vector signal analyzer (VSA) and a block of 8000 sampled symbols was recorded. The performance of each sub-channel was assessed offline by comparing the 8000 de-mapped received symbols with the

Table 4.5 High-speed transmission over standard PMMA SI-POF without equalization

	DR (Gb/s)	POF length, (BER)	Transmitter/receiver	Modulation	DR-distance product (Gb/s·m)
[58]	0.945	100 m ($2 \cdot 10^{-4}$)	LD (8.5 dBm + lensed), PIN-PD (1 mm) + external TIA, adaptive multi-subcarrier modulation	MCM (offline)	94.5
[59]	1–1.5	50 m (10^{-5})	RC-LED—VCSEL, PIN-PD (700 μm) + Gigabit TIA optimization of the system capacity via the bit loading algorithm + error counting	DMT (offline)	50
[54]	1.25	25 m (10^{-9})	LD (14 dBm + collimating lens low launch angle NA = 0.1), PIN-PD (200 μm) + TIA, received optical power −18 dBm	NRZ (high mode filtering)	31.25
[57]	1.0	25 m (10^{-10})	Two LDs (6 dBm), two (PD 1 mm + lens + TIA + post amplifier)	AMGDM + NRZ	25
[60]	1.0	20 m (10^{-8})	LD (655 nm, 6 dBm), integrated linear optical receiver	4-PAM	20
[61]	0.5	40 m (10^{-8})	LD (655 nm, 6 dBm), integrated linear optical receiver	4-PAM	20
[62]	0.4	50 m (10^{-8})	LD (655 nm, 6 dBm), integrated linear optical receiver	4-PAM	20
[56]	1.12	10 m	RC-LED (−3 dBm after 10 m POF), TIA with an integrated PIN-PD (540 μm)	DMT	11.2

transmitted pseudo random binary (PRBS) sequence of length $2^9 - 1$. The resulting average bit error ratio was 10^{-4}.

A comparison between different systems for high-speed transmission over standard PMMA SI–POF without using equalization is introduced in Table 4.5. The data rate-distance product can give a comparison between different systems using different lengths and data rates see Table 4.5. But the system components, power consumption, cost and BER must be taken into account for a detailed comparison. All of the results given in [58, 59] were offline processed and the high value of data rate-distance product cannot be achieved in a real-time system.

The measurements introduced in [54] represent the maximum data rate-distance product in a real-time transmission (31.25 Gb/s·m). But the system needs a high laser optical power (14 dBm) to achieve this result which violates human eye safety. The next higher data rate-distance product is 25 Gb/s·m (achieved by [57]). This system is an expensive one; it needs two lasers and two optical receivers. By using two POF channels the system complexity increases, there is attenuation due to wavelength multiplexing at the transmitter and demultiplexing at the receiver, multiple optics alignment variances, reliability issues associated with individual channel failures and skew management and associated delay. The results in [60–62] show data rate-distance products of 20 Gb/s·m using a low power consumption and a cheap fully integrated optical receiver having a high linearity with 4-PAM modulation format. More details about these integrated optical receivers will be introduced in Chap. 6. Reference [56] introduced an 11.2 Gb/s·m data rate-distance product with a RC-LED using DMT modulation which increases the system complexity, power consumption and costs.

As a conclusion from Table 4.5 for real-time operation the M-PAM system with integrated linear optical receiver results in a better performance and in a simpler system than with the DMT solution. The binary systems can beat the M-PAM one but on the expense of high system costs [57] or large transmitted optical power [54].

4.6.2 By Using Equalization

A binary NRZ signal at a data rate of 1200 Mbit/s over 10 m long SI-POF was reported using a 657nm LD with an optical output power of 7 mW [7]. There is a 2-stage high-pass filter circuit to peak the LD for pre-equalization of the SI-POF's small bandwidth. The optical receiver was a Si-PIN photodiode (Hamamatsu S5052) with an optical lens of 3 mm in diameter. To increase the optical receiver bandwidth despite the large-area photodiode, a 10 Ω optical receiver and 22 V power supply were used. An error-free signal was received for 88 h. Pre-equalizing of the light source (peaking) lowers the light source modulation depth; this reduced the actual power per pulse compared with rectangular pulses without peaking. This is at the expense of system power budget.

Real-time Gigabit Ethernet transmission over up to 25 m Step-Index Polymer Optical Fiber was demonstrated using a commercial 650 nm RC-LED [34]. This is enabled by the use of 4-PAM modulation and adaptive 16-tap receiver equalization including clock recovery, implemented in an FPGA.

The RC-LED driver circuit converts the differential DAC output signal to a suitable current to drive a DC-biased commercially available red RC-LED with a fiber-coupled output power of −1.4 dBm.

The used modulation index is about 60 %. On the receiver side, the optical signal is detected using a PIN photodiode with an active-area diameter of 540 µm with a separate transimpedance amplifier. This signal is post-amplified and filtered using an analogue 5th-order Bessel filter with a 3 dB bandwidth of 440 MHz. Gigabit transmission over 25 m with a resulting BER of $7.4 \cdot 10^{-4}$ and a BER of 10^{-8} for 10 m POF was reported.

In [61] a data rate of 1 Gb/s over 50 m POF was achieved by using an adaptive Decision Feedback Equalizer (DFE) and Forward Error Correction (FEC) which were fully implemented in an FPGA. The measured sensitivities for BER $= 10^{-9}$ were −13.5 dBm using a resonant cavity light emitting diode (RC-LED) and −18.5 dBm with a VCSEL. The optical-to-electrical conversion was then performed by a laboratory receiver (SPD-2 model by Graviton) and a PIN photodiode.

In [43] the transmission part consisted of a current driver, based on an operational amplifier, and an RC-LED from Firecomms. The receiver consists of a Hamamatsu S5971 photodiode, a discrete TIA based on an OPA LMH6624, a TIVCA824 based variable gain amplifier (VGA) and a passive 5th order Butterworth anti-aliasing filter. The receiver part also contained a DC-offset compensation circuit in order to maximize the dynamic range of the ADC. By using this system a data rate of 1 Gb/s over 50 m SI-POF at BER $= 10^{-12}$ was achieved using 16-PAM signaling.
The first integrated POF equalizer was introduced in [11]. A discrete optical receiver (Hamamatsu S5052 800 μm PIN-PD and 28 V reverse bias, Maxim 3266 TIA) was used. The output of the optical receiver was fed to an integrated equalizer chip. A data rate of 1.25 Gbit/s was achieved over 50 m SI-POF using a 650nm RC-LED. The BER was 10^{-6}.
A single chip optical receiver with an integrated equalizer was used to achieve a high performance gigabit transmission over SI-POF [64]. The integrated equalizer can compensate for different POF lengths up to 50 m. The integrated optical receiver is fabricated in a low-cost silicon 0.6 μm BiCMOS technology and has a power consumption of 100 mW. Real-time transmission at data rates of 1.8 Gbit/s over 20 m SI-POF and 1.25 Gbit/s over 50 m SI-POF with high sensitivities and a BER of 10^{-9} was achieved. The optical transmitter was based on an edge emitting laser.
Reference [4] shows an offline processing experiment for 4-PAM 1.25 Gbit/s transmission over SI-POF over up to 75 m using a RC-LED with pre-equalized PAM4 signaling, a Feed-Forward-Equalizer (FFE) and a Decision-Feedback-Equalizer (DFE). 4-PAM at a DR of 1.25 Gbit/s was generated and digitally pre-equalized using an AWG. The pre-equalized sequence drove a RC-LED with an average coupled optical power of -1.4 dBm and optical modulation amplitude (OMA) of 0 dBm. The optical signal was butt-coupled into SI-POF of different lengths. At the receiver side, a PIN-PD with a 540 μm diameter active area was connected to an external TIA with integrated AGC. The received signal was post-amplified and captured afterwards using a digital storage oscilloscope. The stored signals were used to perform simulations in a computer environment. The post equalization was performed offline by FFE or DFE. At fiber lengths of 25, 50 and 75 m the post-equalizer produced BERs of 10^{-5}, $2.1 \cdot 10^{-4}$ and $1.8 \cdot 10^{-2}$, respectively.
A red laser has been used for the giga bit transmitter (665 nm emission, 7 dBm optical power). The other end of the link (50 and 75 m of SI-POF) was coupled to the receiver (a commercial receiver from Graviton, model SPD-2). The received optical power was -11.4 dBm for the 50 m system and -15.2 dBm for the 75 m system [63]. The core of the developed PHY was implemented on an FPGA platform, and it was based on traditional binary on-off modulation at the transmitter and efficient line coding (64B/66B to 65B), keeping the signaling overhead much lower than with e.g., 8B/10B standard Reed-Solomon codes, blind and adaptive DFE at the receiver. The correct operation of the FEC was effective down to -17.5 dBm, which approximately corresponds to a packet error rate of 10^{-7} and a raw bit error rate (before FEC) of approximately 10^{-4}.
Table 4.6 summarizes the work for high-speed transmission over standard PMMA SI-POF using equalization. By comparing the results from Tables 4.5 and 4.6 we can

Table 4.6 High-speed transmission over standard PMMA SI-POF using equalization

	DR (Gb/s)	length, (BER)	Transmitter/receiver	Modulation	DR-distance product (Gb/s·m)
[63]	1	75 m (10^{-7})	LD (665 nm, 7 dBm), graviton optical receiver, model SPD-2, line 64B/66B, adaptive DFE, forward error correction forward error correction	Binary	75
[64]	1.25	50 m (10^{-9})	LD (650 nm, 5 dBm), fully integrated linear optical receiver with an integrated equalizer	Binary	62.5
[4]	1.25	50 m ($2.1 \cdot 10^{-4}$)	RC-LED with −1.4 dBm + pre-distortion filter, PIN-PD (540 um) + external AGC-TIA, received power −9 dBm, post-EQ, adaptive FFE/DFE	4-PAM (Offline)	62.5
[11]	1.25	50 m (10^{-6})	650 nm RC-LED + Hamamatsu S5052 800 μm PIN-PD and 28 V reverse bias, Maxim 3266 TIA + integrated adaptive post-equalizer	Binary	62.5
[45]	1	50 m (10^{-12})	RC-LED, TIA + PD + VGA + post-EQ + coding	16-PAM	50
[65]	1	50 m (10^{-9})	RC-LED/VCSEL + PIN-PD + TIA (SPD-2 by Graviton) + DFE + Forward Error Correction	Binary	50
[33]	1.0	10 m (10^{-8}) 25 m ($7.4 \cdot 10^{-4}$)	RC-LED (−1.4 dBm), TIA with an integrated PIN-PD (540 μm) + discrete post amplifier, post-EQ, adaptive FFE16-tap	4-PAM	25
[7]	1.2	10 m (error-free in 88 h)	LD (8.5 dBm + peaking circuit), PIN-PD + low resistance (10 Ω) amplifier 22 V supply	NRZ	12

conclude that the systems using equalization techniques in Table 4.6 have a better performance than the not equalized ones in Table 4.5. The equalization with binary signaling gives better results than the equalization with M-PAM signaling.

Low data rate transmission over Standard PMMA SI-POF at 10 Mb/s NRZ (Non Return to Zero) data transmission over 425 m SI-POF was presented in [66]. Regarding the optoelectronic part, the proposed system was based on a 520 nm LED and a PIN-PD with integrated TIA (Hamamatsu S6468-02). The −3 dB bandwidth was 35 MHz for the LED and 25.8 MHz for the TIA, while the LED-launched average power was about 1 dBm. The combination of the two optoelectronic elements gives an equivalent cutoff frequency of about 20 MHz, a value that, being significantly higher than the 9 MHz offered by the 425 m POF, does not limit significantly the total available bandwidth. The proposed 425 m system is bandwidth limited mainly

Table 4.7 Low data rate transmission over standard PMMA SI-POF

	DR (Mb/s)	POF length, (BER)	Transmitter/receiver	Modulation	DR-distance product (Gb/s·m)
[6]	100	200 m	Green LED (520 nm, 2.5 dBm), PD + TIA	DMT	20
[67]	100	200 m ($<10^{-8}$) 275 m (10^{-3})	LED (520 nm, +2 dBm, 22 dB attenuation for 275 m POF) + Pre-Equalizer, PD (0.8 mm) + TIA (−23.5 dBm sensitivity BER = 10^{-4}), 3 dB system margin + adaptive post equalizer, RS-FEC (free of error)	8-PAM	20
[68]	170	115 m	LD (655 nm, 6 dBm), integrated linear optical receiver	4-PAM	20

by the fiber, with a bandwidth of 9 MHz, and should cope with at least 35 dB of link attenuation and thus, typically, a 38 dB power budget. RS-FEC was used in order to increase the overall system sensitivity. Before the FEC the BER is 10^{-4} which is smaller than the FEC threshold (10^{-3}); no errors after FEC have been observed in 105 days over 425 m.

100 Mb/s data transmission over 275 m SI-POF was achieved using 8-PAM modulation, RS-FEC coding and pre/post-equalization [67]. The input digital data stream was first sent to a RS-FEC encoder. The digital stream at the output of the RS coder is sent to 8-PAM modulator, then to a pre-emphasis FIR filter. The resulting signal was sent to a block that compensated the LED intrinsic nonlinearity and finally to a digital-to-analog converter (D/A), which directly drove the input of a green LED, giving a peak output power of the order of +2 dBm and a 35 MHz modulation bandwidth. The light was launched to the SI-POF. Considering a target of 275 m distance, for the green wavelength region, the fiber total loss results in about 22 dB. At the receiver side a large-area (800 μm diameter, Hamamatsu S6468-02) photodiode and a linear transimpedance amplifier having a bandwidth of 26 MHz was used. In order to maintain linearity over a large input dynamic range, an AGC system was required. The receiver sensitivity was −23.5 dBm at BER $= 10^{-4}$. The signal was sent to an A/D converter and then to an adaptive equalizer, an 8-PAM demodulator and a FEC decoder. The BER measurements for a pre-coded signal below 10^{-3} showed that, considering a RS-FEC operating threshold (BER $= 10^{-3}$), the maximum reachable distance was 275 m with an optical power margin of 3 dB. A 100 Mb/s free of error transmission over 200 m POF was achieved by using the same system.

A real time DMT system over 200 m SI-POF at 100 Mbit/s upstream and a 100 Mbit/s downstream part was realized based on a VDSL2 chip set in [6]. For the transmission experiment, optoelectronic interfaces based on 520 nm LEDs and large-area photodiodes were built. Due to the lack of optical SI-POF couplers, a bidirectional link comprising of two pieces of SI-POF (200 m each) was set up. The average transmitted optical power was 2.5 dBm and the SI-POF attenuation was approximately 83 dB/km at a wavelength of 520 nm. Using this configuration, a data rate of 101.832 Mbit/s upstream and 100.384 Mbit/s downstream was achieved. First test results showed an average latency of around 1.1 ms and 0 % packet loss for all packet sizes. If the MII (Media Independent Interface) would allow the sum of the up- and downstream, data rates can be transmitted in both directions, resulting in a total data rate of 202 Mbit/s for a single SI-POF of 200 m.

A comparison between different systems for low data rate transmission over standard PMMA SI-POF is introduced in Table 4.7. All of the results in Table 4.7 have a data rate-distance product 20 Gb/s·m. The used modulation format in [6] is DMT. DMT modulation increases the system complexity, power consumption and the total costs. A simpler modulation format (8-PAM) was used in [67]. But pre- and post-equalization were needed and RS-FEC was performed. This increases the system complexity and cost. The result in [68] shows a data rate-distance products of 20 Gb/s·m using a low power consumption, cheap fully integrated optical receiver having a high linearity with 4-PAM modulation format. There was no need neither for pre/post-equalization nor RS-FEC.

Chapter 5
Integrated Photodiode

The integrated photodiode is the main component and the first stage in the integrated optical receiver. The photodiode converts the optical power into an electrical current. The photodiode should convert photons into charge carrier pairs with maximum efficiency and to transport them to the electrodes as fast as possible. Also the capacitance of the photodiode must be as low as possible to keep the high frequency loss of the photocurrent low to reach highest possible sensitivity. These demands are partially conflicting. So a trade-off is necessary, especially due to the wavelength dependence of the penetration depth. More details about the theory of integrated photodiodes can be found in [69, 70].

5.1 Optical Absorption

Light travels with the velocity c in a medium with the optical refractive index n,

$$c = \frac{C_0}{n} \tag{5.1}$$

where c_0 as the velocity of light in vacuum.
Photons are characterized by their wavelength in vacuum λ_0, which can be calculated from the medium independent frequency f:

$$\lambda_0 = \frac{C_0}{f} \tag{5.2}$$

Every single photon owns its energy E:

$$E = hf = \frac{hC_0}{\lambda_0} \tag{5.3}$$

M. Atef and H. Zimmermann, *Optical Communication over Plastic Optical Fibers*, Springer Series in Optical Sciences 172, DOI: 10.1007/978-3-642-30388-3_5,

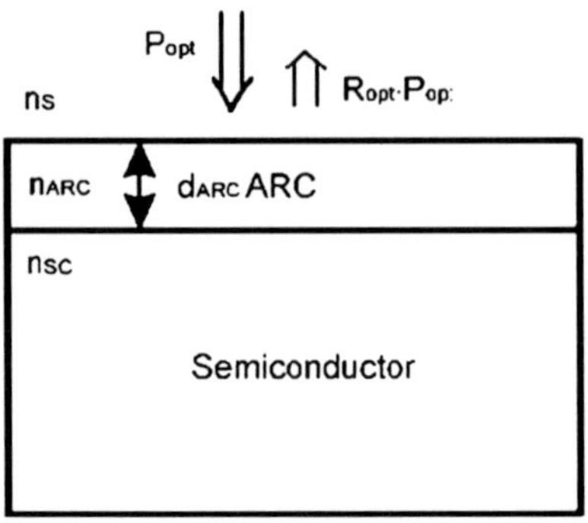

Fig. 5.1 Reflection on the semiconductor surface with antireflection coating

Photons will produce electron-hole pairs in silicon when the photon energy E is larger than the band gap energy ($Eg = 1.1\,\text{eV}$ for silicon). For silicon the critical wavelength is $\lambda_c = 1.1\,\mu\text{m}$, above this value the silicon is transparent and there is no absorption. So silicon photodiodes are usable for wavelengths from near infrared up to the ultraviolet wavelength range.
A part of the total optical power incident to the photodiode (P_{OPT}) is reflected from the surface. The optical power P_0 directly below the surface in the semiconductor can be calculated with the reflection factor R_{OPT} by:

$$P_0 = (1 - R_{OPT})P_{OPT} \tag{5.4}$$

The reflection coefficient can be minimized with an anti-reflection coating (ARC) at the surface of the semiconductor for a certain wavelength or a more or less small wavelength range. In Fig. 5.1 the principle structure of the antireflection coating and the distribution of the optical power below the semiconductor surface can be seen. The optimum values of the thickness of the ARC d_{ARC} and its refractive index n_{ARC} can be calculated with the formulas (5.5) and (5.6) with the refractive index of the surrounding n_S and the refractive index of the semiconductor n_{SC}. The materials SiO_2 ($n_{ARC} = 1.45$) and Si_3N_4 ($n_{ARC} = 2.0$) are most adequate for the ARC coating.

$$n_{ARC} = \sqrt{n_s n_{sc}} \tag{5.5}$$

$$d_{ARC} = \frac{\lambda_0}{4n_{ARC}} \tag{5.6}$$

The incident photons with wavelengths smaller than the critical wavelength λ_c have enough energy to transfer an electron to the valence band. This process absorbs the photon and an electron-hole pair is generated. The optical absorption coefficient α defines the penetration depth $1/\alpha$ of the light into the semiconductor in accordance to Lambert-Beer's law:

$$P(x) = P_0(x)e^{-\alpha x} \tag{5.7}$$

The penetration depth $1/\alpha$ for 650 nm is 3.3 µm and the relative absorbed power length for 90 % ($d_{90\,\%}$) of the total power is 7.6 um.
The generation rate per volume $G(x)$ can be expressed by:

$$G(x) = \frac{P(x) - P(x + \Delta x)}{\Delta x} \cdot \frac{1}{Ahf} \tag{5.8}$$

which results in:

$$G(x) = \frac{\alpha P_0}{Ahf} e^{-\alpha x} \tag{5.9}$$

5.2 Photodiode Responsivity

The conversion factor from optical power P_{OPT} to electrical current is defined by the responsivity R, it can be calculated by:

$$R = \frac{I_{PH}}{P_{OPT}} = \frac{q\lambda_0\eta}{hc} \tag{5.10}$$

The responsivity is an important link between the optical power and the current. It is necessary to evaluate the sensitivity (minimum optical power for data transmission for a certain bit-error ratio) of an optical receiver.

5.3 Quantum Efficiency

The quantum efficiency gives the percentage of generated electron-hole pairs per incident photons, which are transported to the electrodes. It can be calculated by another form of relation (5.10) with:

$$\eta = \frac{Rhc_0}{q\lambda} \tag{5.11}$$

The quantum efficiency (5.12) can be divided into optical quantum efficiency η_o and an internal quantum efficiency η_i. The optical quantum efficiency accounts for the partial reflection of the optical power at the surface of the semiconductor. It can be maximized for a certain wavelength range by introducing an ARC layer at the surface of the photodiode

$$\eta = \eta_i \eta_0 \tag{5.12}$$

$$\eta_0 = (1 - R_{OPT}) \tag{5.13}$$

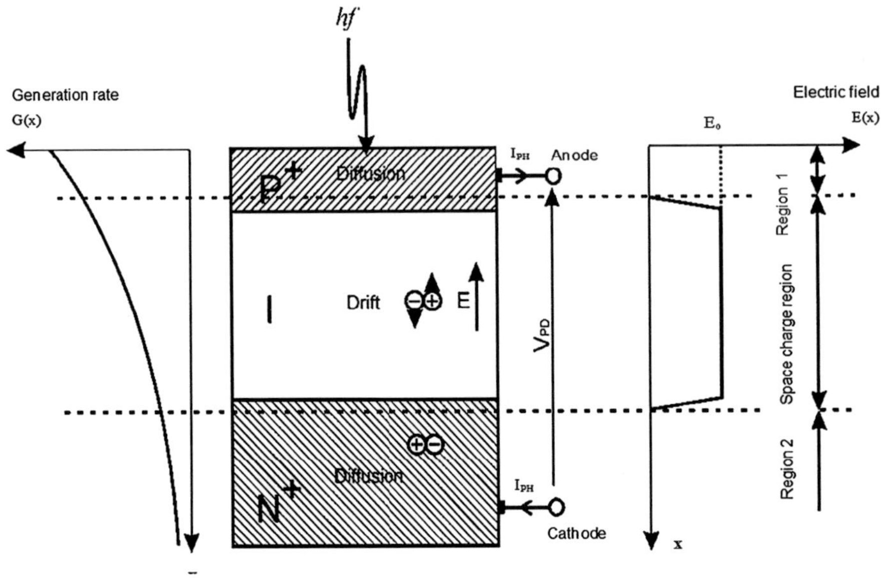

Fig. 5.2 Structure of a PIN photodiode

The internal quantum efficiency η_i defines how effectively the optical power is converted to electron-hole pairs, which contributes to the photocurrent at the connection terminals. η_i has a frequency dependent behavior, since two transport effects are involved, carrier drift and carrier diffusion.

5.4 Carrier Drift and Diffusion

Carrier drift is a fast process and carrier diffusion is a slow process. Carrier drift occurs when an electric field is present. Then the charge carriers can reach a high velocity. This is a wanted effect since it improves the speed of the photodetector. In contrast, carrier diffusion occurs when no electric field is present and movement of the charge carriers is slow, which reduces the bandwidth of the whole detector. This means for a fast photodiode, most of the charge carriers must be photo-generated in a zone where a strong electric field exists. The PIN-diode structure (Fig. 5.2) fulfills this requirement best, since it offers a thick zone with a high nearly constant electric field, where charge carriers are generated.
Generally spoken for short optical wavelengths (e.g. $\lambda = 410\,\text{nm}$) surface effects are dominant, and for long optical wavelengths (e.g. $\lambda = 850\,\text{nm}$) charge carriers are generated deep in the silicon, where no electric field is present. Diffusion should be avoided where possible, since this effect causes the photodiode to slow down. Often a photodiode must be optimized for a particular wavelength to suppress carrier

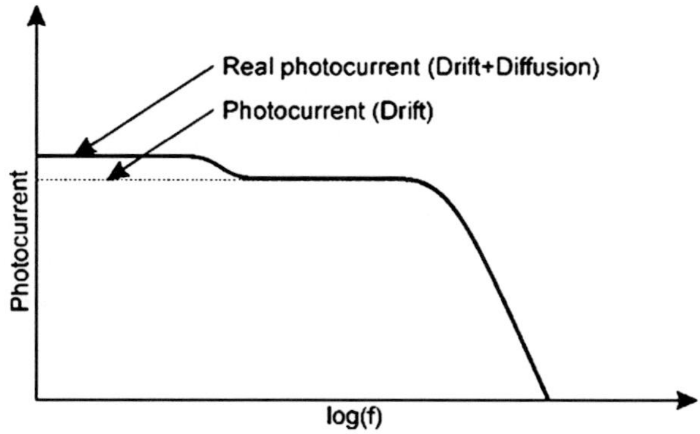

Fig. 5.3 Frequency response of the photocurrent with drift and diffusion

diffusion. A very thick intrinsic layer could be used, but a large voltage is necessary to ensure that the intrinsic region is completely depleted. For short wavelengths (e.g. $\lambda = 410\,\text{nm}$) diffusion in the surface electrode can dominate, which should then be thin to reduce carrier diffusion (Fig. 5.3).

5.5 PIN Photodiode Capacitance

The capacitance of a PIN photodiode can be calculated for the case when the intrinsic region is fully depleted. This is the case for the normal operating condition, since only under this circumstance high speed can be reached. The capacitance can then approximately be calculated with (5.14) , which is the formula for a plate capacitor, neglecting the boundary capacitance. Silicon acts as the dielectric with a relative permittivity of $\varepsilon_r = 11.9$.

$$C = A \cdot \frac{\varepsilon_r \varepsilon_0}{d_I} \tag{5.14}$$

5.6 Photodiode Speed

Depending on the specific application, the thickness of the intrinsic zone of the photodiode must be optimized. For a good trade-off, the main parameters that must be known are the maximum data rate, the operating wavelength, type of photodiode and the maximum voltage available for the photodiode. A main characteristic of integrated optical receivers is the achievable sensitivity, which depends strongly on the responsivity of the photodiode and also the capacitance, so minimum capacitance and maximum responsivity lead to maximum sensitivity. This could be achieved by increasing the thickness of the intrinsic layer. But this contradicts to speed, since

increasing the intrinsic layer thickness without increasing the reverse voltage would lead to a double penalty (quadratic dependence) since the electric field decreases and so the speed of the charge carriers decreases. Additionally the distance the charge carriers have to drift increases and so the whole transition time of the charge carriers increases dramatically. To solve this problem partially the voltage across the diode has to be increased according to the increase of the thickness of the intrinsic layer. The main goal is that the electric field is at a high constant level where the charge carriers travel with saturation drift velocity. In practice, however, the saturation velocity is not achieved in most cases due to limited available supply voltage.
Often a certain data rate is given, so the photodiode has to achieve a certain rise- and fall-time. In a first order approximation the necessary 10–90 % rise- and fall-time of the photodiode can be calculated from the formula (5.15) with t_d as the drift time through the intrinsic zone.

$$\frac{t_r + t_f}{2} = \frac{1}{3DR} \approx t_d \tag{5.15}$$

When rise- and fall-time are nearly equal, then the formula (5.15) simplifies to formula (5.16) and gives approximately the drift-time of the charge carriers through the drift zone. In both cases the factor 3 indicates a conservative approximation and for an aggressive approximation the factor can be two.

$$t_r = t_f = \frac{1}{3DR} \approx t_d \tag{5.16}$$

The cut-off frequency of the PIN-photodiode can be estimated by:

$$f_c \approx \frac{2.4}{\pi(t_r + t_f)} \tag{5.17}$$

Under the assumption that the drift velocity reaches saturation ($v_s = 10^7$ cm/s) the maximum allowable thickness of the intrinsic layer can easily be calculated by:

$$d_I \approx t_d v_s \tag{5.18}$$

The necessary voltage for the PIN-diode can now also be calculated with the electric field where saturation occurs.

$$V_{sat} = E_{sat} d_I \tag{5.19}$$

5.7 Structure of the Implemented Photodiode

Considerably better results can be obtained with vertical PIN photodiodes than with other PD structures [71–73]. The insertion of an intrinsic layer in PIN-PD assures a thick depletion region even at low reverse voltages. If an appropriate thickness of

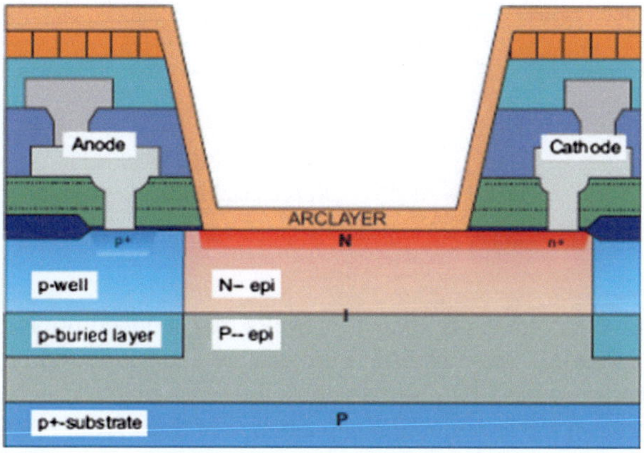

Fig. 5.4 Vertical PIN photodiode in BiCMOS technology

the intrinsic layer is chosen for the given optical wavelength, nearly all carriers are photo-generated within the depletion region and slow carrier diffusion is avoided. There have been several attempts to integrate photodiodes in a BiCMOS process. In [74] a PIN photodiode with a bandwidth of 700 MHz is reported. Since its intrinsic layer is formed by a standard N-well between a P+ source/drain region and an N+ buried layer it has a capacitance of $C_D = 0.32\,\mathrm{fF/\mu m^2}$ and a responsivity of only $R = 0.07$ A/W at a wavelength of 850 nm. A PIN photodiode integrated in a standard-buried-collector (SBC) based BiCMOS process achieved a bandwidth of 300 MHz at a reverse voltage of $V_R = 3$ V and a responsivity of $R = 0.35$ A/W at a wavelength of 780 nm [75]. A buried N^+ cathode (first additional mask) and a P-type buried isolation layer (second additional mask) had to be implanted. Both lead to auto-doping problems during the following N^- epitaxial for the intrinsic zone of the PIN photodiode [75]. Furthermore a two-step N^- epitaxial procedure was necessary to combine PIN photodiode and transistors in the same process. A third mask was necessary to increase the doping in the collector of the bipolar transistors to avoid the Kirk effect [76]. All together in [75], therefore, a considerable additional process complexity was needed to integrate the PIN photodiode excluding a low-cost performance.

The PIN-PD integrated with the optical receivers described in this book uses the approach introduced in [71–73]. It is possible to buy the P^-/P^+ epitaxial wafers with a P^- doping concentration of about $10^{13}\,\mathrm{cm^{-3}}$ from wafer suppliers. The 0.6 μm BiCMOS process used provides a mask for defining the optical window with antireflection coating and a mask for blocking out the wells. The native N-type epitaxial layer, which exists in regions where no P-well or N-well is generated, is part of our PIN photodiode. Reducing the doping concentration of this N-type epitaxial layer to approximately $10^{14}\,\mathrm{cm^{-3}}$ was the only process modification needed. A vertical PIN photodiode is created by placing an N^+source/drain region in an area excluded from automatic P-well generation. This results in a thick almost intrinsic layer reaching from the N^+ cathode to the P^+ substrate, Fig. 5.4.

Table 5.1 Measured results for PIN photodiodes in 0.6 μm BiCMOS technology obtained with an optical wavelength of 660 nm

Responsivity	0.512 A/W
Quantum efficiency	96.4 %
Rise time	516 ps
Fall time	545 ps
Calculated 3 dB bandwidth	720 MHz
Area capacitance	0.0158 fF/μm^2

Because all active elements are placed in wells, the reduced doping concentration of the N-type epitaxial layer has little influence on the MOS and bipolar transistor parameters. Therefore the PIN photodiode of the work in [71–73] is much easier to integrate. Because of the low junction capacitance of this vertical PIN photodiode, it is possible to design high-speed OEICs with photodiodes having large diameters. The described method of implementing a PIN photodiode in an OPTO ASIC BiCMOS process does not require any additional process modification and delivers excellent reproducible results.

Table 5.1 lists measured results of the integrated PIN photodiode. The antireflection coating is optimized for red light. The quantum efficiency of 96 % was measured for wavelengths of 660 nm. The rise and fall times of 516 and 545ps resulted at a wavelength of 660 nm. The thick depletion region reduced the area capacitance to 0.0158 fF/μm^2.

There is no statistically significant difference between the parameters of transistors in the standard and in the PIN process. This can be explained by the fact that all these transistors are placed in wells.

Chapter 6
Transimpedance Amplifier

6.1 Introduction

Photodiodes do not use any amplification effect and have responsivity (ratio of output current to input light intensity) that is for example $\leq$0.5 A/W for a silicon photodiode at a light wavelength of 650 nm. Therefore a combination of a photodiode with an amplifier is necessary in all practical applications. In optical receivers current-to-voltage converters are necessary in order to convert the photocurrent delivered by the photodiode into an output voltage which is proportional to the input current. The transimpedance amplifier is the most suitable preamplifier configuration used in optoelectronic receivers. For most optical receiver applications these amplifiers need a high and also variable gain, high bandwidth, low noise and low input impedance. The main characteristics of transimpedance amplifiers are discussed in [69].

6.2 Transimpedance Amplifier

The preamplifier is used to convert the incoming photocurrent into an output voltage, which is amplified by the following stages. The simplest way to do this conversion is a resistor between the PD output and the supply voltage as shown in Fig. 6.1. The preamplifier is one of the determining parts concerning the sensitivity and bandwidth of an optical receiver. The sensitivity mainly depends on the responsivity of the PD and the input referred noise current of the circuit. Due to the fact that the output current of the PD is the smallest signal in the circuit, this point is the most sensitive concerning noise. The signal-to-noise ratio is most critical at the input node of the preamplifier. Therefore, the noise of the preamplifier is the dominating part of the input referred noise current. Again the noise of the resistor R and the first amplifying stage are the deciding factors.
The most interesting characteristics of the preamplifier, therefore, are the bandwidth and the input referred noise of the circuit. For the simple receiver shown in Fig. 6.1

M. Atef and H. Zimmermann, *Optical Communication over Plastic Optical Fibers*, Springer Series in Optical Sciences 172, DOI: 10.1007/978-3-642-30388-3_6,

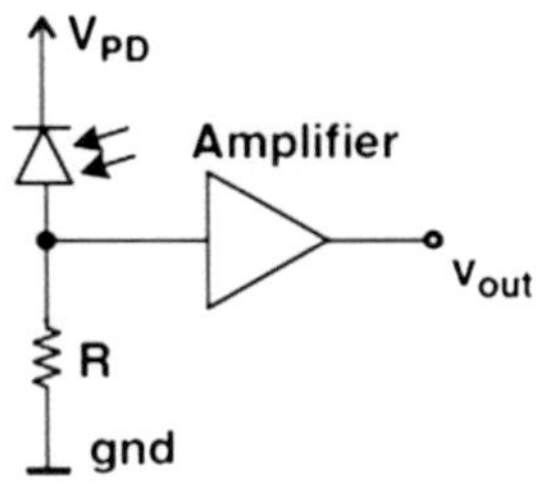

Fig. 6.1 Simplest possibility of the preamplifier

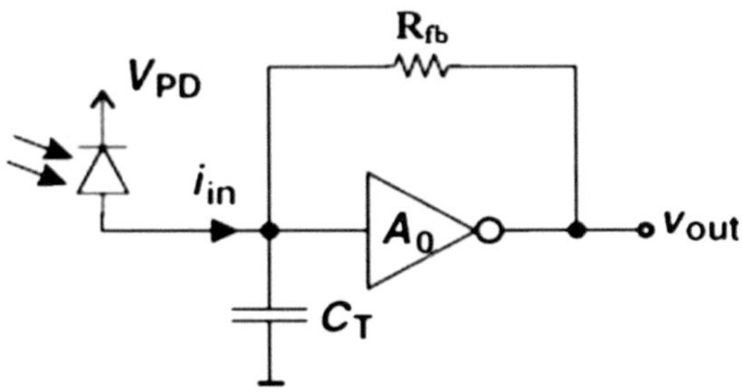

Fig. 6.2 Basic circuit of a TIA as preamplifier

the bandwidth is indirectly related to the capacitance of the input node and the resistor R. In this simple model, the capacitance of the input node consists of the capacitance of the PD, and the input capacitance of the amplifier. To achieve a high bandwidth therefore the resistance R has to be small, as well as the capacitance of the input node.

The noise of the circuit shown in Fig. 6.1 depends also on the resistor R, the capacitance of the input node and the first amplifier stage of the following amplifier. To achieve high bandwidths the resistor R must be small and therefore its noise current dominates the sensitivity of the optical receiver. With a more complicated circuit, for example a TIA, a better performance can be achieved.

6.3 Shunt-Shunt Feedback TIA

Figure 6.2 shows the basic circuit of a shunt feedback TIA as a preamplifier. In the TIA circuit C_T consists of the capacitance of the PD and the input capacitance of the TIA.

Let us compare the small-signal transfer functions of the two structures in Figs. 6.1 and 6.2, which is the transimpedance gain v_o/i_{in}.

From Fig. 6.1

$$\frac{v_o}{i_o} = \frac{R}{1 + j\omega C_T R} \tag{6.1}$$

From Fig. 6.2

$$\frac{v_o}{i_o} = \frac{R_{fb}\frac{A_o}{A_o+1}}{1 + j\omega R_{fb} C_T \frac{A_o}{A_o+1}} \tag{6.2}$$

The DC transimpedance gain for the simple TIA in Fig. 6.1 is the transimpedance resistor value R_{fb}. For high amplifier gains A_o this is also the case for Fig. 6.2. Comparing the frequency behavior, the dominant pole which defines the small-signal bandwidth for the two circuits are defined by:
From Fig. 6.1

$$BW = \frac{1}{2\pi C_T R} \tag{6.3}$$

From Fig. 6.2

$$BW = \frac{A_o + 1}{2\pi C_T R_{fb}} \tag{6.4}$$

Equation (6.3) shows, that the bandwidth of the simple resistor TIA is completely defined by a given transimpedance and photodiode capacitance. Equation (6.4) shows, that the shunt-shunt feedback TIA has an approximately A_o times higher bandwidth compared to a simple resistor TIA. The amplifier gain A_o is design dependent and higher than 1 for frequencies up to the amplifiers transit frequency f_t. Therefore the increased bandwidth for the same transimpedance value is one of the most important advantages of a shunt-shunt feedback TIA compared to a simple resistor solution.
The advantage of a shunt-shunt feedback TIA compared to the simple circuit described before is the fact that the bandwidth is indirectly related to the resistor R_{fb} divided by the open-loop gain A_o of the *TIA* (R_{fb}/A_o). Therefore the noise can be decreased for a given bandwidth, because of a large resistor R_F. The smaller C_T and the larger R_{fb}, the higher the sensitivity is for a given responsivity (more discussion will be presented in the next sections). For a high bandwidth it is important to have small C_T and small R_{fb}. C_T is dominated by the PD capacitance, a tradeoff between sensitivity and bandwidth has to be found for the feedback resistor.

6.3.1 Frequency Response

The operational amplifier is characterized by its frequency dependent amplification

$$A(\omega) = \frac{A_o}{1 + j\frac{\omega}{\omega_T} A_o} \tag{6.5}$$

Application of first and second Kirchhoff's law onto the circuit shown in Fig. 6.3 gives:

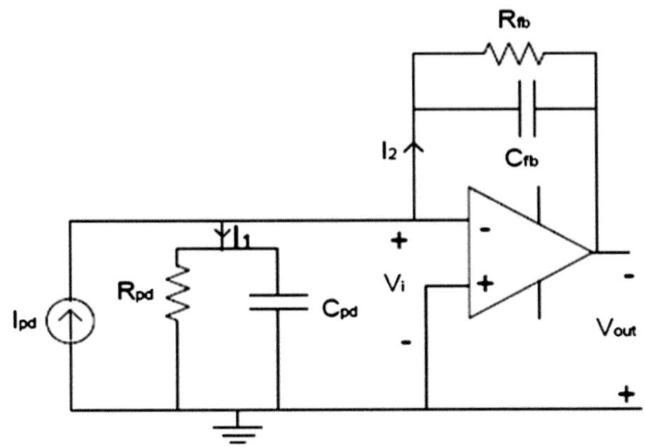

Fig. 6.3 Transimpedance amplifier circuit with photodiode model

$$V_i + I_2 Z_{fb} + V_o = 0 \tag{6.6}$$

where $Z_{fb} = R_{fb}||C_{fb}$, and

$$I_{ph} + I_1 - I_2 = 0 \tag{6.7}$$

Inserting $I_1 = \frac{V_i}{Z_{pd}}$, where $Z_{pd} = R_{pd}||C_{pd}$, and $V_i = \frac{V_o}{A(\omega)}$ into (6.6) and (6.7) leads to:

$$I_{pd} + \frac{V_o}{\mathrm{A}(\omega) \cdot Z_{pd}} - I_2 = 0 \tag{6.8}$$

and

$$V_o\left(1 + \frac{1}{A(\omega)}\right) + I_2 Z_{fb} = 0 \tag{6.9}$$

Multiplying (6.8) by Z_{fb} and adding it to formula (6.9) results in the transfer function $G(\omega)$
which is the transimpedance and possesses the dimension Ω:

$$G(\omega) = \frac{V_o}{I_{pd}} = -\frac{Z_{fb}}{1 + \frac{1}{A(\omega)}\left(1 + \frac{Z_{fb}}{Z_{pd}}\right)} \tag{6.10}$$

with $A(\omega) = \frac{A_o}{1+j\frac{\omega}{\omega_T}A_o}$ Eq. (6.10) can be written as [69]:

$$G(\omega) = \frac{V_o}{I_{pd}}$$
$$= -\frac{R_{fb}}{1 - R_{fb}(C_{pd} + C_{fb})\left(\frac{\omega^2}{\omega_T}\right) + j\omega\left(\frac{1}{\omega_T} + \frac{R_{fb}}{R_{pd}\omega_T} + R_{fb}C_{fb} + R_{fb}\frac{C_{pd}}{A_o}\right)} \quad (6.11)$$

With the characteristic angular frequency ω_N

$$\omega_N = \sqrt{\frac{\omega_T}{R_{fb}(C_{pd} + C_{fb})}} \quad (6.12)$$

and the damping factor D

$$D = 0.5.\sqrt{\frac{\omega_T}{R_{fb}(C_{pd} + C_{fb})}} \cdot \left(\frac{1}{\omega_T} + \frac{R_{fb}}{R_{pd}\omega_T} + R_{fb}C_{fb} + R_{fb}\frac{C_{pd}}{A_o}\right) \quad (6.13)$$

formula (6.11) simplifies to:

$$G(\omega) = -\frac{\omega_N^2 \cdot R_{fb}}{\omega_N^2 + j2\omega D\omega_N - \omega^2} \quad (6.14)$$

The magnitude of the transfer function can be determined from formula (6.14) to be:

$$|G(\omega)| = -\frac{\omega_N^2 \cdot R_{fb}}{\sqrt{\omega_N^4 - 2\omega_N^2\omega^2 + \omega^4 + 4\omega^2 D^2 \omega_N^2}} \quad (6.15)$$

In Fig. 6.4 the normalized magnitude of the transfer function is plotted for different values of the damping factor D. Equation (6.15) can have a resonance (gain-peak) at $\omega_{GP} = \sqrt{\frac{\omega_N}{1-2D^2}}$, provided that the damping factor D < 0.7. The value of the normalized transfer function at the gain-peak frequency is:

$$\frac{|G(\omega)|}{R_{fb}} = -\frac{1}{2D\sqrt{1 - D^2}} \quad (6.16)$$

In transimpedance amplifiers a feedback capacitance C_{fb} can be used to control the damping factor D. Gain-peaking is a possibility to increase the bandwidth of the transimpedance amplifier at the cost of a higher group delay variation and longer settling time due to a higher overshoot in the step response [69].
For an ideal operational amplifier with A_o and ω_T equal to ∞ the gain in (6.11) becomes:

$$G(\omega) = -\frac{R_{fb}}{1 + j\omega(R_{fb}C_{fb})} \quad (6.17)$$

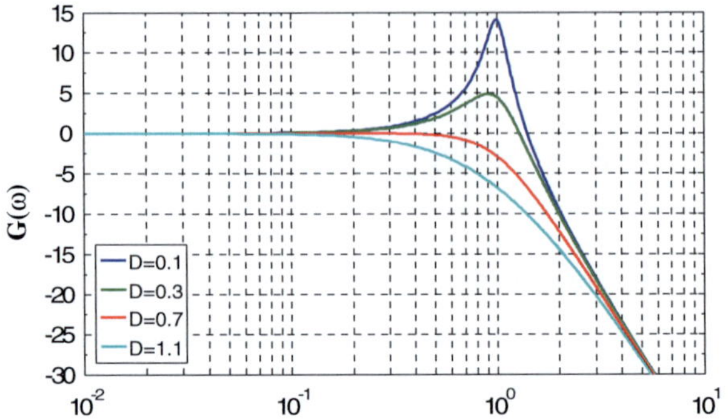

Fig. 6.4 Normalized magnitude of the transfer function G(ω) for different values of the damping factor D

and the −3 dB bandwidth can be calculated from the following equation:

$$BW = \frac{1}{2\pi \cdot (R_{fb} C_{fb})} \tag{6.18}$$

6.3.2 TIA Noise Analysis

In this section noise models of transimpedance amplifiers will be presented. First a TIA with an ideal amplifier will be discussed. Two different TIAs with BJT input stage and MOSFET input stage are described in detail subsequently. The input referred noise current of the TIA depends on the input node capacitance C_T, the feedback resistor R_{fb}, and the transistor noise of the amplifier's input stage.

Ideal-Amplifier TIA

First the general model of the ideal TIA is considered. In Fig. 6.5 the noise sources of the individual TIA components are included. The noise of the TIA depends on the technology and topology of the circuit. As a first estimation we define a very simple circuit which also converts an input current into an output voltage, a pin photodiode and transimpedance amplifier. In the easiest case this amplifier is a common-emitter or a common-source circuit, respectively. With more complicated designs better results can be achieved, but as an upper bound for noise we consider the circuit shown in Fig. 6.6. The main noise sources are the thermal noise of the resistor R_F and the equivalent input noise sources of the amplifier $\overline{i^2_{n.amp}}$ and $\overline{v^2_{n.amp}}$. We can calculate the total equivalent input noise current spectral density as follows [69]:

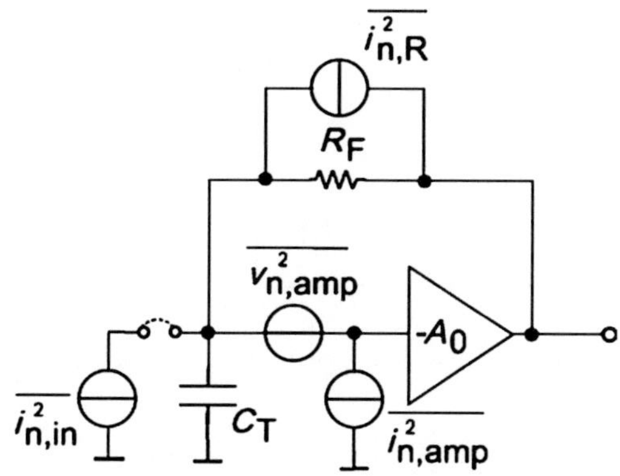

Fig. 6.5 Basic circuit of a TIA including general noise sources

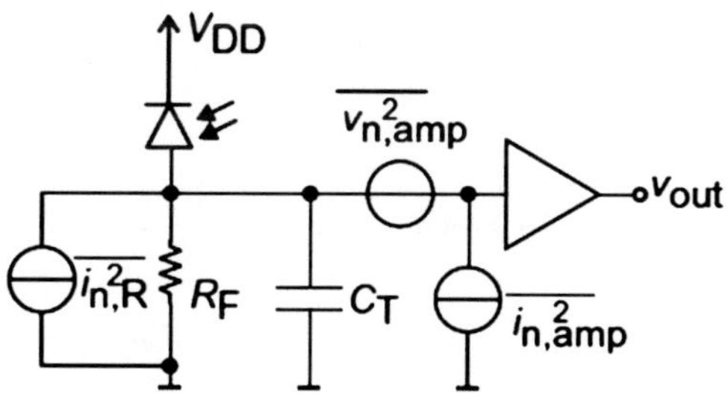

Fig. 6.6 PIN photodiode with amplifier for upper bound for noise analysis

$$\overline{i^2_{n.in}} = \overline{i^2_{n.amp}} + \frac{\overline{v^2_{n.amp}}}{\left|\frac{R_F}{1+j\omega C_T R_F}\right|} + \overline{i^2_{n.R}} \tag{6.19}$$

which leads to:

$$\overline{i^2_{n.in}} = \overline{i^2_{n.amp}} + \overline{v^2_{n.amp}}\left(\frac{(1 + 4\,\pi^2\,f^2 C_T^2 R_F^2)}{R_F^2}\right) + \overline{i^2_{n.R}} \tag{6.20}$$

Equation (6.20) shows that the equivalent input noise current spectral density raises proportional to C_T^2 and f^2 at high frequencies. The total input noise current is obtained by integrating Eq. (6.20) over the bandwidth.

6.3.3 TIA with Common-Emitter Input Stage

For bipolar and heterojunction bipolar common-emitter (CE) input stages we assume the circuit shown in Fig. 6.7. The noise sources of the amplifier of this circuit are

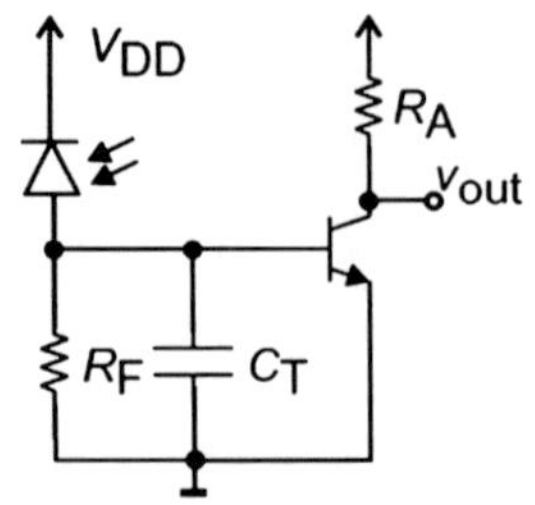

Fig. 6.7 PIN-BIP amplifier circuit for small-signal bipolar-TIA noise analysis

equal to the equivalent input noise sources of the bipolar transistor $\overline{v_{n.Q}^2}$ and $\overline{i_{n.Q}^2}$. The noise of the load resistor R_A is attenuated by the gain of the bipolar transistor when referred to the input and therefore at first approximation neglected.
The equivalent input noise voltage of the bipolar transistor can be calculated by using the following equation [70]:

$$\frac{\overline{v_{n.Q}^2}}{\Delta f} = 4K_BT\left(r_b + \frac{1}{2g_m}\right) \tag{6.21}$$

where K_B is Boltzmann's constant, and the equivalent input noise current density of the bipolar transistor can be calculated from:

$$\frac{\overline{i_{n.Q}^2}}{\Delta f} = 2q\left(I_B + \frac{K_1}{2q}\frac{I_B^a}{f} + \frac{I_c}{|\beta(jf)|^2}\right) \tag{6.22}$$

Equation (6.20) leads with the equivalent input noise sources of the bipolar transistors, see (6.21) and (6.22) to:

$$\frac{\overline{i_{n.in}^2}}{\Delta f} = \frac{4K_BT}{R_F} + 2q\left(\frac{I_C}{\beta} + \frac{K_1}{2q^2}\left(\frac{I_C}{\beta}\right)^a + \frac{I_C}{\left|\frac{\beta}{1+j\frac{f}{f_\beta}}\right|^2}\right) + \frac{4(1+2\pi f C_T R_F)^2 \cdot K_BT\left(r_b + \frac{1}{2g_m}\right)}{R_F} \tag{6.23}$$

For a TIA with BIP input stage, a high g_m decreases the noise at high frequencies. But for the low frequency range the noise even increases with a high collector current (high g_m), due to the higher base current. For noise optimization the input transistor of a bipolar TIA needs therefore a low base resistance r_b, high g_m and a high current gain β to reduce the base current I_B.

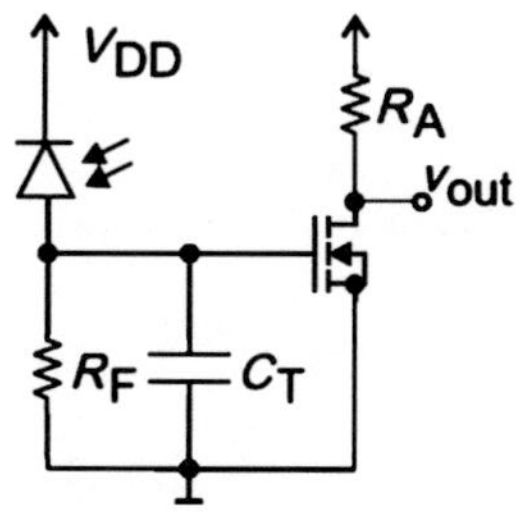

Fig. 6.8 PIN-FET amplifier circuit for small-signal FET-TIA noise analysis

6.3.4 TIA with Common-Source Input Stage

For a MOSFET common-source (CS) input stage the amplifier circuit is shown in Fig. 6.8. Again the noise sources of the amplifier equal the equivalent input noise sources of the FET $\overline{v_{n.M}^2}$ and $\overline{i_{n.M}^2}$. The noise of the load resistor R_A again is neglected, because it is attenuated by the transconductance of the MOSFET, if it is referred to the input.
The equivalent input noise voltage of the MOSFET can be calculated by using the following equation [70]:

$$\frac{\overline{v_{n.M}^2}}{\Delta f} = 4K_B T \frac{2}{3g_m} + K_f \frac{I_D^a}{g_m^2 f} \tag{6.24}$$

and the equivalent input noise current density of the *MOSFET* transistor can be calculated from:

$$\frac{\overline{i_{n.M}^2}}{\Delta f} = \omega^2 C_{gs}^2 \left(4K_B T \frac{2}{3g_m} + K_f \frac{I_D^a}{g_m^2 f}\right) \tag{6.25}$$

$$\begin{aligned}\frac{\overline{i_{n.in}^2}}{\Delta f} = {} & \frac{4K_B T}{R_F} + \frac{4\pi^2 f^2 C_T^2}{g_m^2}\left(4K_B T \frac{2}{3g_m} + K_f \frac{I_D^a}{g_m^2 f}\right) \\ & + \frac{1 + 4\pi^2 f^2 C_T^2 R_f^2}{R_f^2 g_m^2}\left(4K_B T \frac{2}{3g_m} + K_f \frac{I_D^a}{g_m^2 f}\right) + 2qI_D\end{aligned} \tag{6.26}$$

Substituting the noise generators shown in (6.24) and (6.25) leads to:
For low-noise applications, the g_m of the input MOSFET and therefore its drain current I_D should be as high as possible.
Dependent on the frequency and the photodiode capacitance the FET front-end can have a performance advantage compared to a bipolar front-end. Generally, for lower data rate the FET front-end can achieve lower noise, see Eqs. (6.23) and (6.26).

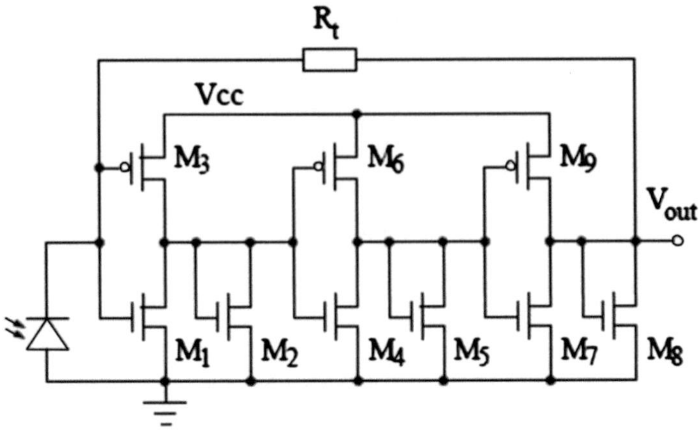

Fig. 6.9 Circuitry of three-stage inverter based CMOS transimpedance amplifier

This is, since the BJT generates a higher equivalent input noise current; due to the base current and the base series resistor. For BJT transistors with high transit-frequency f_T it is possible that the BJT front-end reaches significantly better performance than the FET-frontend at high frequency. Lower noise can especially be expected for the BJT when the source (photodiode) impedance is low. This is the case at high frequencies and/or large photodiode capacitances.

6.3.5 Multistage Inverter Based CMOS TIA

The circuit shown in Fig. 6.9 is a single-ended TIA in pure CMOS technology, which means a low-cost implementation [77]. This circuit uses a 3-stage inverter based amplifier to achieve a high gain. The output of the 3rd stage is connected to the TIA input via the transimpedance resistor R_t. Every stage of the three amplifier stages consists of 3 MOSFET transistors (like the first stage with M_1, M_2 and M_3). The gain of every stage is defined by the transconductance of the three transistors, as for the first stage:

$$A_1 = \frac{g_{m,M_1} + g_{m,M_3}}{g_{m,M_2} + g_{ds,M_1} + g_{ds,M_2} + g_{ds,M_3}} \approx \frac{g_{m,M_1} + g_{m,M_3}}{g_{m,M_2}} \tag{6.27}$$

This equation shows that the gain is given by a g_m ratio, and is therefore mostly defined by transistor matching. A significant advantage of this three-stage circuit is the high amplifier gain, which means also a high TIA bandwidth according to Eq. (6.4). Another advantage is the low supply voltage the circuit needs. The main disadvantage of this circuit is the sensitivity to power supply noise. Another critical design issue is the stability behavior due to the multi-stage amplifier feedback.

A noise cancelling technique can be used with the multistage CMOS inverter based TIA [78] to enhance the sensitivity of the conventional multistage inverter based amplifier.

6.4 Common-Base/Gate Input Stage

Another TIA implementation is shown in Fig. 6.10 [79]. This TIA is implemented in a BiCMOS technology. Instead of using a common-emitter amplifier, this circuit uses a common-base (CB) input stage, defined by transistor Q_1. The output of the common-base amplifier is connected to the actual TIA input. This uses a common-source stage, with the NMOS transistor M_1 as amplifier, the BJT transistors Q_2 and Q_3 as output buffers and the resistor R_F as transimpedance. The new pole introduced by the common-base stage should be placed sufficiently high such that it does not interfere with the frequency response of the shunt-feedback TIA. The low input resistance of the common-base stage, which is about l/g_m, helps to satisfy this condition.

$$Z_{in} \approx 1/g_{m,Q_1} \tag{6.28}$$

$$\omega_{-3\,\mathrm{dB}} = \frac{1 + A_{V,M_1}}{R_F[C_{gs,M_1} + (1 + A_{V,M_1})C_{gd,M_1} + C_{\mu,Q_1}]} \tag{6.29}$$

$$A_{V,M_1} = g_{m,M_1} \cdot R_D \tag{6.30}$$

The collector bias current of BJT Q_1 is lower than the MOSFET drain bias current for the same transconductance (g_m) and hence reduces the supply voltage requirements.

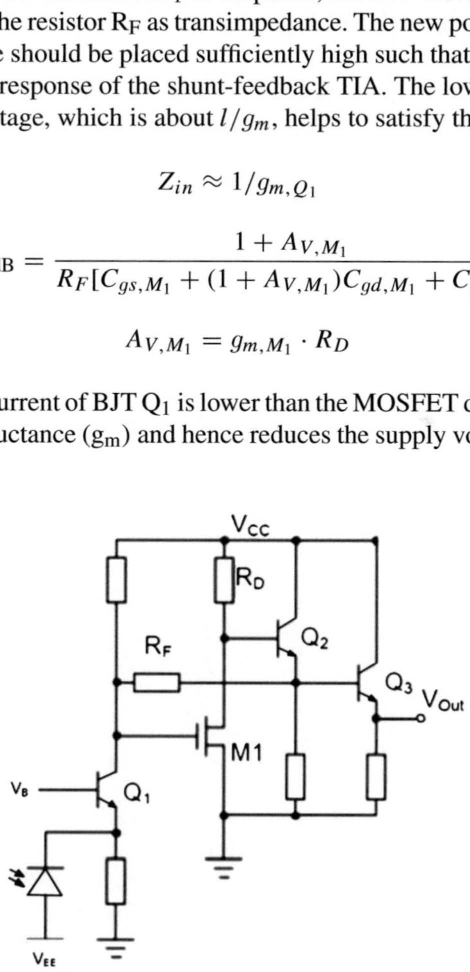

Fig. 6.10 Transimpedance amplifier using a common-base input stage

The input noise current of MOSFET (M_1) is less than that of a BJT, and the optimum MOSFET size can be selected. Also, the MOSFET provides better isolation between stages.
The main advantage of this circuit is that the capacitance of the input node is decoupled from the TIA input by the common-base stage. Therefore the TIA bandwidth can be optimized, since it is independent of the photodiode capacitance, which can be large and vary for different applications. Compared to the CS-TIA topology, the CB-TIA has an improved speed performance. On the otherhand, the common-base input stage is an additional significant noise source, which is the main drawback of this circuit. Also the current consumption of this architecture will be higher, due to the additional input stage.

6.5 Regulated-Cascode TIA

The regulated cascode (RGC) has very low input impedance which can support a high bandwidth at low power consumption [80]. The circuit structure using the RGC configuration can isolate the capacitance effect. Figure 6.11 shows the circuitry of a conventional RGC-TIA. The transimpedance gain is given by:

$$Z_{T,RGC}(0) = R_1 \tag{6.31}$$

and the input impedance can be calculated from:

$$Z_{in,RGC}(0) = \frac{1}{g_{m,n1} \cdot \{1 + g_{m,n2} \cdot (R_2 // r_{ds,n2})\}} = \frac{1}{g_{m,n1} \cdot (1 + A_{cs})} \tag{6.32}$$

where $(1 + A_{cs})$ is the gain of the local feedback loop of the CS stage (M_{n2}, R_2) and the CS voltage gain is:

$$A_{cs} = g_{m,n2} \cdot (R_2 // r_{ds,n2}) \tag{6.33}$$

The main advantage of the RGC-TIA is its low input impedance which can isolate the effect of larger photodiode capacitance. If it is assumed again that the dominant pole is located at the input, the bandwidth of this topology is given by:

$$BW = \frac{g_{m,n1} \cdot (1 + A_{cs})}{2\pi C_{in}} = \frac{g_{m,n1} \cdot (1 + g_{m,n2} \cdot R_2 // r_{ds,n2})}{2\pi C_{in}} \tag{6.34}$$

Compared to the bandwidth of the common-gate (or common-base) stages, the RGC bandwidth can be increased for the same transconductance of M_1. For the same bandwidth a CG-stage needs a higher g_{m1} than the RGC-stage which means a larger current is needed through M_1. The power consumption of the RGC-TIA is lower than that of the CG-TIA.
When the shunt-shunt feedback TIA consumes the same power as the CG-TIA and RGC-TIA, the transconductance of the input transistor of the voltage amplifier is

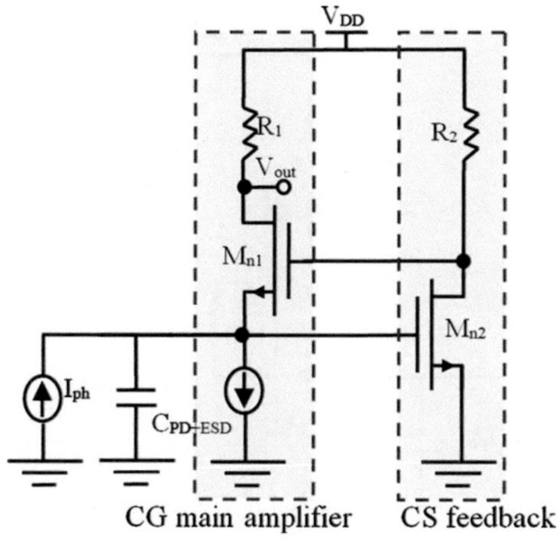

Fig. 6.11 Regulated cascode transimpedance amplifier

increased and input noise current of the shunt-shunt feedback TIA decreases. For the same power consumption, the shunt-shunt feedback TIA has superior noise performance compared with the regulated-cascode TIA.

The transimpedance gain is limited by R_1 which cannot be increased due to limited low supply voltage. The input impedance can be decreased by increasing the CS voltage gain A_{cs}. But this gain is limited also by the low supply voltage. Increasing the current through M_2 will increase $g_{m,n2}$ but $r_{ds,n2}$ will decrease. The maximum current is also limited by

$$(V_{DD} - V_{GS,n1} - V_{GS,n2})/R_2 \approx (V_{DD} - 2 \cdot V_{TH})/R_2 \tag{6.35}$$

Increasing the current will be on the expense of a smaller R_2 value. As a result the A_{cs} value will be limited. The RGC-TIA is difficult to be implemented at a low supply voltage. A modification in the RGC-TIA was introduced in [81, 82] for low voltage operation.

6.6 Inverter Based Common-Drain Feedback TIA

A TIA employs an inverter with an active common-drain feedback (ICDF-TIA) to achieve a higher bandwidth at lower power consumption than with the normal CS-TIA and a better sensitivity than with the conventional RGC-TIA introduced in [83]. Figure 6.12 shows the circuitry of the proposed ICDF-TIA where PMOS transistors replace R_1 and R_2. The inverter stage with M_{n2} and M_{p2} has a voltage gain

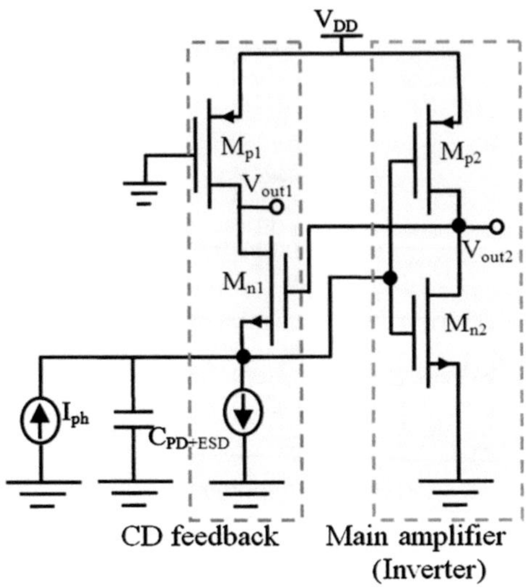

Fig. 6.12 Inverter based common-drain feedback TIA [83]

$$A_{inv} = (g_{m,n2} + g_{m,p2}) \cdot (r_{ds,n2} // r_{ds,p2}) \tag{6.36}$$
$$= g_{m2,eff} \cdot (r_{ds,n2} // r_{ds,p2}) \tag{6.37}$$

where

$$g_{m2,eff} = g_{m,n2} + g_{m,p2} \tag{6.38}$$

The input impedance becomes:

$$Z_{in,CSCG}(0) = \frac{1}{g_{m,n1}[1 + (g_{m,n2} + g_{m,p2}) \cdot (r_{ds,n2} // r_{ds,p2})]} \tag{6.39}$$
$$= \frac{1}{g_{m,n1}[1 + A_{inv}]} \tag{6.40}$$

The input impedance of the proposed ICDF-TIA can be decreased by increasing the sum of $g_{m,n2}$ and $g_{m,p2}$.
The ICDF-TIA's input impedance can be lower than the RGC-TIA's input impedance due to the increase of the effective g_m by adding the inverter stage (M_{n2} and M_{p2}) which in general has a higher gain than the common-source amplifier stage (M_{n2} and R_2). The smaller input impedance of the ICDF-TIA has the effect of increasing the TIA bandwidth at the same current consumption as for the RGC-TIA. By using the ICDF-TIA the same bandwidth can be obtained at lower current consumption.
The main difference between the RGC and the ICDF TIAs is the output node and which stage is the main amplifier or the feedback. The RGC output Fig. 6.11 is taken from the drain of the CG amplifier (M_{n1}, R_1). The common-source amplifier (M_{n2}, R_2) is working as local feedback stage. Whereas the ICDF output node V_{out2}

(Fig. 6.12) is the drain of the inverter amplifier (M_{n2}, M_{p2}). M_{n1} and M_{p1} are working as local feedback stage. The feedback input is the gate of M_{n1} and the feedback output is the source of M_{n1}. M_{n1} and M_{p1} work as a common-drain feedback (CDF) for the main amplifier (the inverter M_{n2}, M_{p2}). The ICDF-TIA transimpedance gain with respect to the output V_{out2} can be calculated from the small-signal model:

$$Z_{T,CSCG}(0) = \frac{(r_{ds,n1} + r_{ds,p1}) \cdot (r_{ds,n2} // r_{ds,p2}) \cdot g_{m2,eff}}{1 + g_{m,n1} \cdot r_{ds,n1} + g_{m,n1} \cdot r_{ds,n1} \cdot (r_{ds,n2} // r_{ds,p2}) \cdot g_{m2,eff}} \quad (6.41)$$

$$= \frac{(r_{ds,n1} + r_{ds,p1}) \cdot A_{inv}}{1 + g_{m,n1} \cdot r_{ds,n1} \cdot (1 + A_{inv})}$$

$$= \frac{1}{g_{m,n1}[1 + A_{inv}]} \quad (6.42)$$

The ICDF-TIA's transimpedance gain is larger than that of the RGC-TIA for the same power consumption. The actual output voltage node of the proposed TIA is V_{out2} because of its higher gain compared to V_{out1}. The output V_{out1} has a transimpedance gain of $r_{ds,p1}$ which is smaller than the transimpedance gain of V_{out2}. The multi-threshold technology enables to use V_{out2} with higher gain and enough voltage headroom (M_{n2} and M_{p2} are selected to be low-threshold transistors). By comparing simulations for V_{out1} and V_{out2} in the ICDF-TIA, a higher transimpedance gain results for using V_{out2} [83].

6.7 TIA with Gain Control

The dynamic range of a TIA is defined by its overload current and its sensitivity. For a shunt-feedback TIA, both quantities are related to the value of the feedback resistor, and thus the dynamic range can be extended by making this resistor variable with the input optical signal power.

The variable feedback resistor can be implemented with a MOSFET operating in the linear region, connected in parallel to a fixed resistor to improve the linearity and to limit the maximum resistance (Fig. 6.13). The automatic gain control (AGC) can be implemented with a circuit that determines the output signal strength (peak detector or average power) to control the gate voltage of the MOSFET. The AGC output is a DC signal with a value depending on the input signal.

The stability of the AGC-TIA is an important issue. When the feedback resistance (R_F) varies while keeping the open-loop gain fixed, both the bandwidth and the quality factor will change. If R_F is reduced, the low-frequency pole at $1/(R_{fb} \cdot C_T)$ increases, which may lead to peaking at a given fixed loop gain. The bandwidth of the AGC-TIA tends to increase with the magnitude of the input signal.

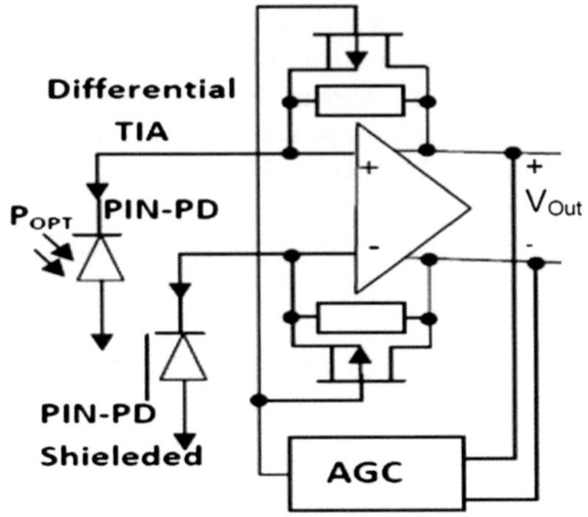

Fig. 6.13 Transimpedance amplifier with gain control

6.8 Post Amplifier

There are two different post amplifiers: limiting and automatic gain control (AGC) amplifier. The main amplifiers have differential inputs and outputs to amplify a small input voltage signal coming from the TIA to an output signal which is sufficient for the reliable operation of the clock and data recovery.

6.8.1 Limiting Amplifier

Amplifiers normally show a linear transfer function for input signals below critical input signal amplitude. For input signals above the critical input signal amplitude there appear nonlinearities. The transfer function of the limiting amplifier (LA) is depicted in Fig. 6.15a [84]. The region of linear operation and the area of dominating nonlinearities can be seen clearly. Due to the given power supply voltage the limiting effect of the LA appears naturally and no special design is necessary. Nevertheless, limiting effects like pulse-width distortion and delay variations must be minimized by a controlled limiting function.

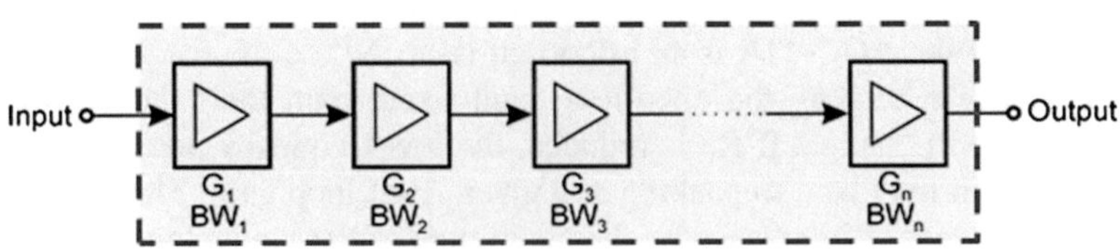

Fig. 6.14 Chain of bandwidth-limited amplifiers

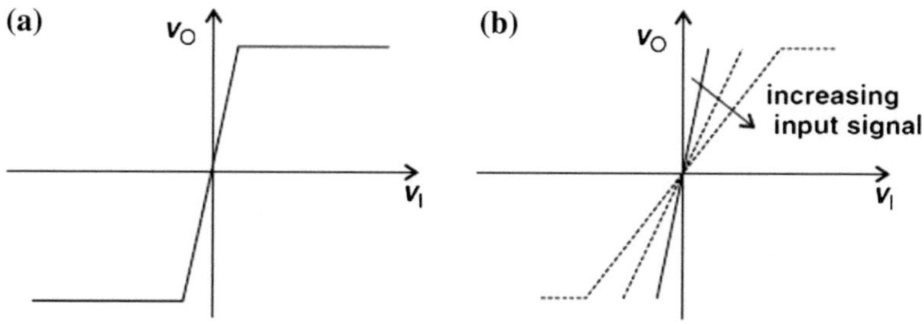

Fig. 6.15 DC transfer characteristics of (**a**) LA; (**b**) AGC amplifier [84]

The total bandwidth must be sufficient to amplify the data signal without ISI. The bandwidth should be limited to the lowest possible value to achieve optimum noise performance. In Fig. 6.14 a chain of amplifying blocks with their basic properties of gain G_i and bandwidth BW_i is shown. An amplifier chain with certain total gain A_{vTotal} with a certain total bandwidth BW_{Total} must be implemented, with a minimum number of stages n.
For an amplifier with n identical stages (same gain G_i and same bandwidth BW_i) the total gain G_{Total} of the amplifier chain can be calculated with relation (6.43), and the total bandwidth BW_{Total} can be calculated from relation (6.44).

$$G_{Total} = [G_i]^n \tag{6.43}$$

$$BW_{Total} = BW_i \sqrt{2^{1/n} - 1} \tag{6.44}$$

6.8.2 AGC Amplifier

AGC amplifiers consisting of a variable gain amplifier (VGA) and an AGC circuit that keeps the output swing constant over a wide range of input swings. Figure 6.15b shows the basic function of an automatic gain control amplifier (AGC). For large input signals the gain is reduced to keep the AGC in the linear region. This reduction is only possible in a defined region. Input signals above this region are limited too.
Because the LA does not have a gain control circuit, it is easier to be designed than AGC amplifiers. Furthermore, its power dissipation, bandwidth and noise are superior to AGC amplifiers.
The linear transfer function of the AGC amplifier preserves the signal wave forms and permits analog signal processing to be performed on the output signal (such as multilevel signaling). Limiting amplifiers when operated in the limiting region, however, cause much of the information in the input signal to be lost.

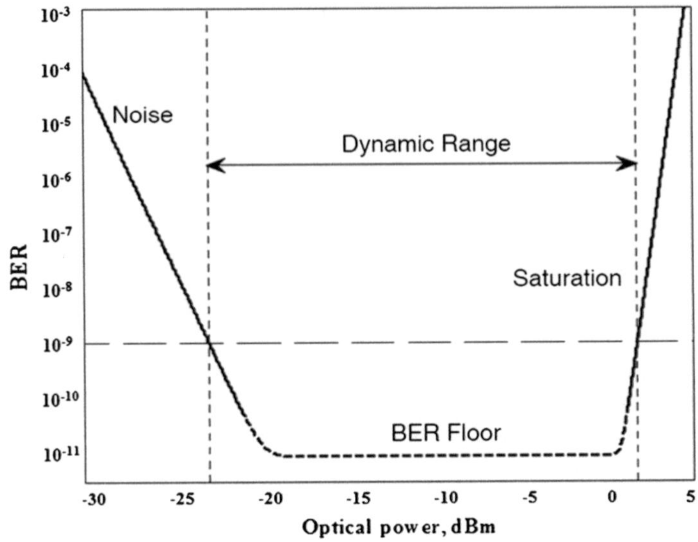

Fig. 6.16 BER as a function of the optical power defining the dynamic range region and its limits

6.9 Dynamic Range

One important feature of the TIA is its dynamic range (DR). The dynamic range is the working range for the TIA for the input optical power going from its minimum (sensitivity at certain BER) to its maximum input optical power (overload optical power), which is limited by TIA saturation.

The dynamic range can be calculated as the ratio between the maximum and minimum input optical power:

$$DR = 10log\frac{P_{max}}{P_{min}} \text{ dB} \tag{6.45}$$

The dynamic range can be increased by increasing the sensitivity (TIA with lower noise) or increasing the maximum received input optical power where the TIA reaches saturation. The maximum limit can be increased by using a TIA with AGC or gain compression stage. Figure 6.16 shows the BER as a function of the optical input power defining the dynamic range region and its limits.

Chapter 7
Integrated Optical Receiver and Multilevel Transmission over PMMA SI-POF

Optical absorption in a semiconductor material is the key-effect to convert optical power into an electrical current. This conversion is done by an integrated photodiode in the integrated optical receivers introduced in this book. Discrete photodiodes can be connected to transimpedance amplifier ICs in multi-chip packages by wire bonding or in hybrid integrated circuits. The advantage of discrete solution is that the performance for each device can be optimized separately by using special photodiode and IC technologies. The drawback is high costs due to a complex production and assembly process. Other disadvantages of using discrete photodiodes are potential noise sources and bandwidth reduction due to bonding pads and package pins.
An alternative low-cost solution in the visible spectral range is the integration of photodiodes in standard CMOS and BiCMOS technologies. The reason is mainly the maturity and performance of standard IC technologies. Therefore optoelectronics integrated circuits (OEIC) are a competitive solution for many low-cost consumer and sensor applications. Another advantage of an OEIC is that there are no bonding pads, no bond wires and no package pin parasitics between photodiode and amplifier input which gives a better performance than with discrete solutions.

7.1 Optical Receiver with 622 MHz Bandwidth

7.1.1 Circuit Description

The block diagram of the optical receiver front-end is shown in Fig. 7.1. The first element in the optical receiver is the photodiode (PIN-PD). There are two identical PIN-PD one to receive the optical signal and the other is shielded to balance the differential transimpedance amplifier (TIA). This improves the power supply rejection ratio very efficiently. The large diameter of the PIN-PD (300 μm) makes the use of large core diameter POF easier.

M. Atef and H. Zimmermann, *Optical Communication over Plastic Optical Fibers*, Springer Series in Optical Sciences 172, DOI: 10.1007/978-3-642-30388-3_7,

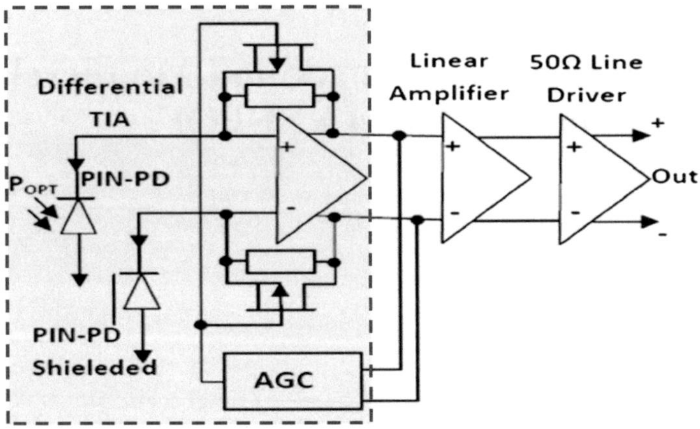

Fig. 7.1 Block diagram of the optical receiver

The PIN-PD is followed by a TIA which converts the photocurrent coming from the PD into a voltage. An automatic gain control circuit (AGC) controls the gain of the TIA in order to avoid the overloading of the TIA and to increase the dynamic range of the optical receiver. This wide dynamic range for the optical receiver allows using it to receive multilevel signals [85, 86]. The use of multilevel signals instead of binary signals allows the use of an optical receiver with a smaller bandwidth and relaxes the frequency constraints on the circuit design, on the other hand the circuit complexity increases since the receiver must deliver a linear response over a large dynamic range. This leads to a more sophisticated design of the AGC compared to the conventional receiver, where a significant part of the large dynamic range is realized with simple limiting amplifiers. So, the conventional limiting receiver for binary data will not amplify the different symbol levels equally. As a result, the voltage levels for the output signal will not be equally spaced. This fact will make the signal detection and decoding more difficult or even impossible. However, the wide range linear AGC-receiver has variable gain and always operates in the linear region. As a result, each level of the output multi-level signal will be amplified by the same gain.

The complete front-end is designed for a symbol rate of 1.25 Gbaud and a maximum overall transimpedance of approximately 40 kΩ. The front end is supplied by a single 3.3 V supply voltage. The last stage in the front-end, which forms the interface with the following off-chip signal processing units, is a 50 Ω output impedance driver (Fig. 7.2). The micrograph of the fabricated optical receiver with the two integrated PDs is shown in Fig. 7.3.

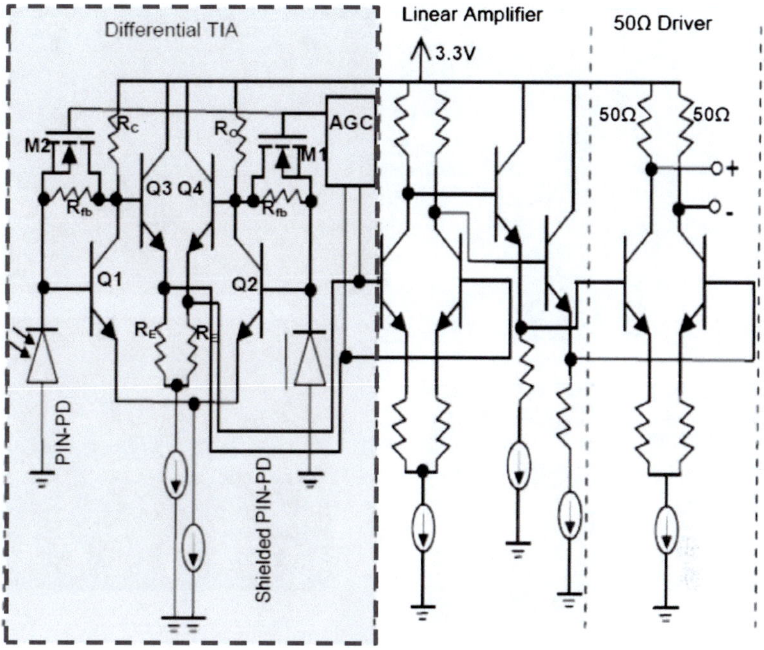

Fig. 7.2 Circuit diagram of the integrated optical receiver [87]

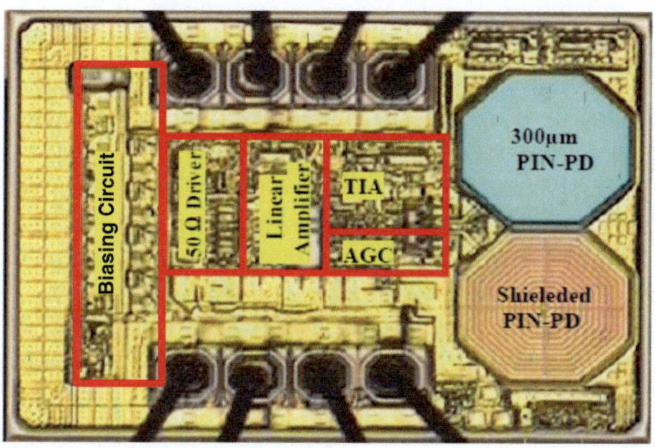

Fig. 7.3 Microphotograph of the optical receiver with the integrated photodiodes

7.1.2 Experimental Results

The optical receiver's frequency response was measured with a network analyzer (HP 8753E). A calibration was performed to cancel the effect of the laser source and the connected measurement cables on the receiver frequency response. Photos of the

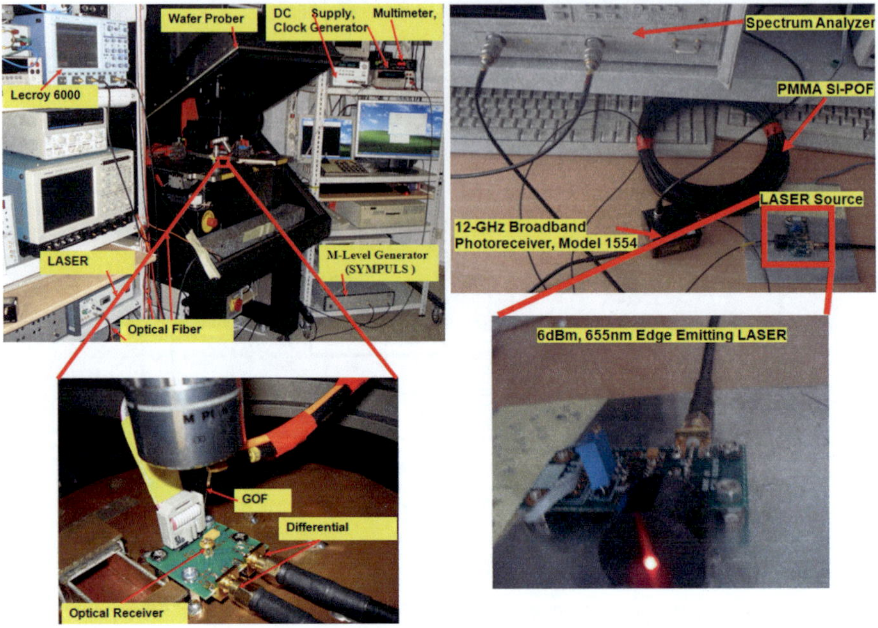

Fig. 7.4 Photos of the sensitivity and frequency measurement setup

sensitivity and frequency measurement setup are shown in Fig. 7.4. The measured receiver's −3 dB bandwidth is 622 MHz (Fig. 7.5). With the 622 MHz bandwidth a data rate of 1.25 Gbit/s can be received using binary signals. The data rate can be doubled to 2.5 Gbit/s by using 4-PAM signaling at the same bandwidth.

Sensitivity measurements were carried out at 660 nm wavelength. For these measurements a red VCSEL was modulated by a pseudo-random bit sequence (PRBS) with a length of $2^{31} - 1$. A communication signal analyzer was used to analyze the output signal and to determine the bit error rate of the optical receiver.

Figure. 7.6 shows the measured eye diagram of the optical receiver using binary data (two levels) transmission at a bit rate of 1.25 Gb/s at a wavelength of 660 nm and an average optical power of 4 μW. For a bit error rate of 10^{-9} the measured sensitivity was −24 dBm.

In the beginning of our multilevel experiments the 4-PAM signal is generated using the method introduced in Sect. 2.6.1, Figs. 2.6 and 2.7. Due to the bad source eye diagram generated by this method the transmitted data rate was limited to 1.4 Gbit/s. The measured output voltage of the optical receiver is constant and equal to 750 mV for any input optical power of the highest level larger than 20 μW. This results in equally spaced levels (e.g. 250 mV for $M = 4$) due to the long time constant (about 100 μs) of the AGC and enables the correct detection of the multilevel signal at constant threshold voltages.

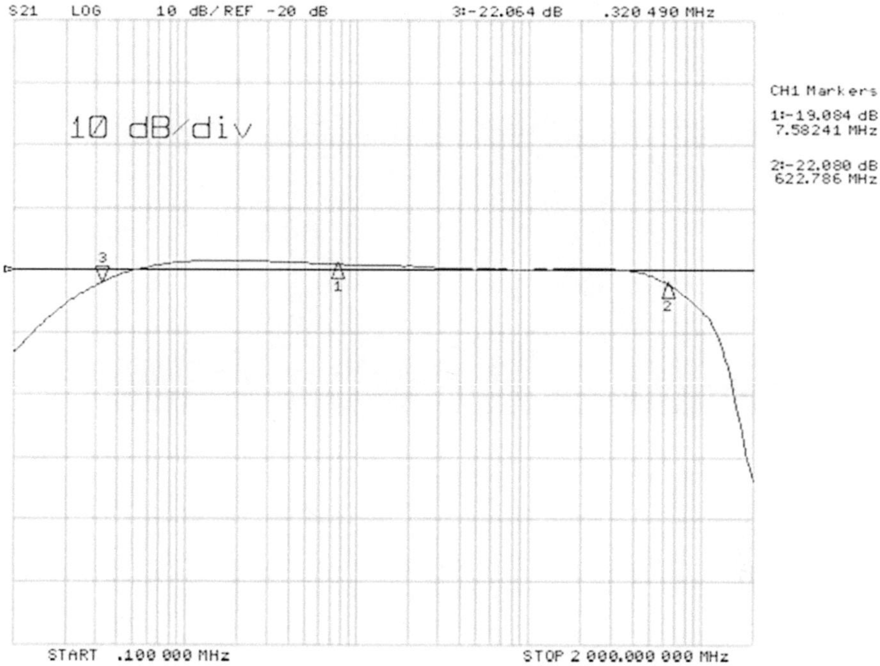

Fig. 7.5 Measured frequency response of the optical receiver

Figure 7.7 shows the measured output eye diagram of the optical receiver using a 4-PAM signal at DR = 1.4 Gbps, PRBS $2^{31} - 1$, a wavelength of 675 nm and an average optical power of −11 dBm. The eye diagram for the 4-PAM signal shows an equal eye opening for each level. For sensitivity measurements a fast 670 nm laser diode was modulated by 4-PAM with a PRBS length of $2^{7} - 1$ at different optical power levels. Another sensitivity measurement was made with a PRBS length of $2^{31} - 1$. A communication signal analyzer (CSA8000) was used to analyze the output signal eye diagram, measure the mean voltage level and the noise variance for each level which was used to calculate the SER from relation (2.22).

The SER for DR = 1, 1.4 Gbps and PRBS = $2^{7} - 1$ as well as $2^{31} - 1$ is calculated after laser extinction ratio correction from 3.5 to 10 was made using Eq. (4.6). The proposed optical receiver reaches sensitivities of −14.8 and −11.9 dBm for DR of 1 and 1.4 Gb/s, respectively, at a wavelength of 675 nm with a BER of 10^{-9} and a PRBS length of $2^{7} - 1$. With a PRBS length of $2^{31} - 1$ and under the same conditions, sensitivities of −12.1 and −11.2 dBm for DR of 1 and 1.4 Gb/s were measured [87]. Another experiment was performed with a better 4-PAM generator introduced in Figs. 2.10, 2.11 and 2.12 [88]. The 4-PAM signal with a DR of 1 Gbit/s (500 Mbaud) was generated on a personal computer by MATLAB software then transmitted by a 12 bit arbitrary waveform generator card (Acquitek DA11000-16M) introduced in

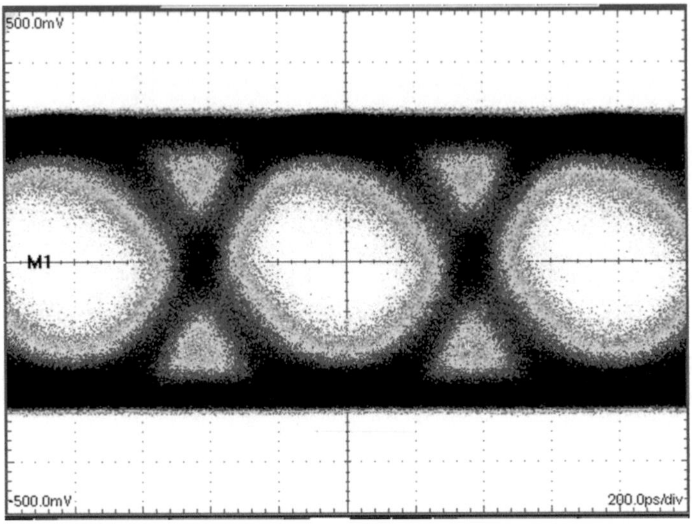

Fig. 7.6 Measured eye diagram of the frond-end optical receiver for a binary signal at a bit rate of 1.25 Gb/s, PRBS $= 2^{31} - 1$ and an average optical power of a 4 μW [86]

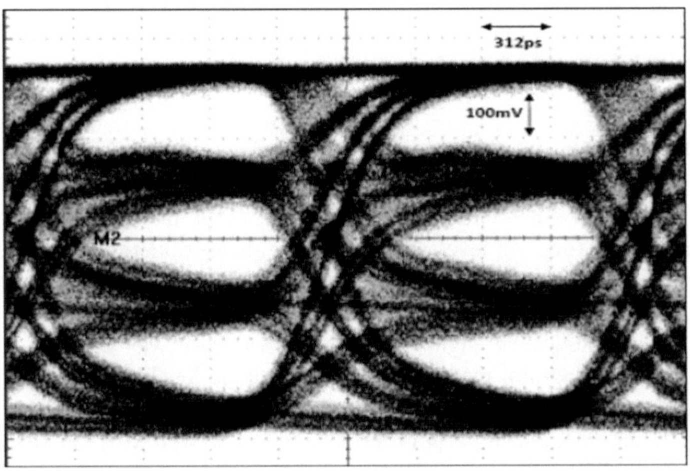

Fig. 7.7 Measured eye diagram of the optical receiver 4-PAM signal (PRBS $= 2^{31} - 1$, 1.4 Gbps and $P_{opt} = -11$ dBm) [87]

Fig. 2.10. The experimental work was evaluated at a DR of 1 Gbit/s (500 Mbaud) because of the limited sampling rate of the used arbitrary waveform generator.

The 1 Gbit/s 4-PAM signal was used to modulate a 670 nm VCSEL with an extinction ratio ER $= 5$. The optical signal was transmitted through the fiber and detected by the

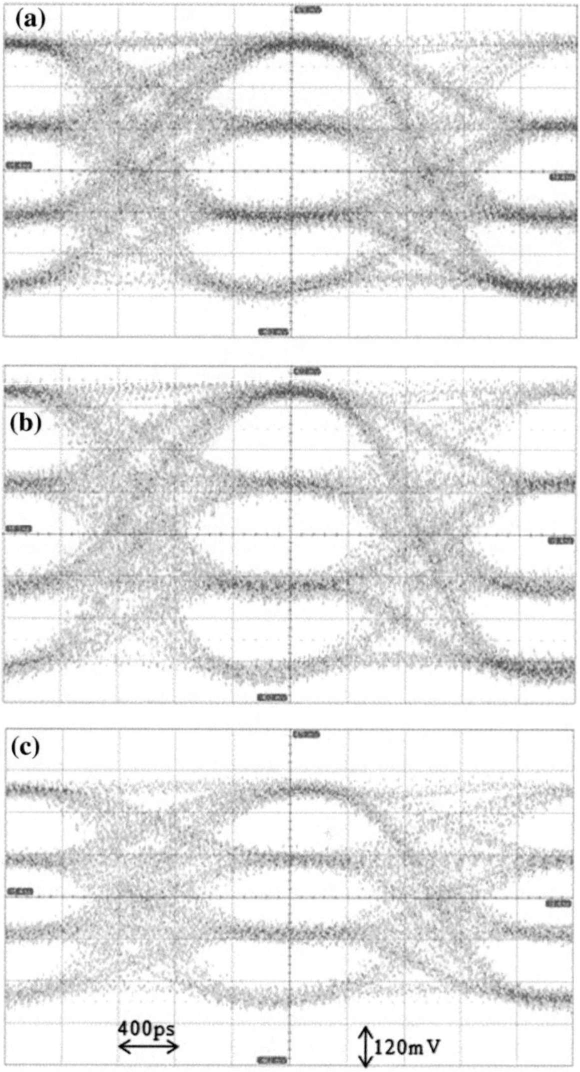

Fig. 7.8 Measured eye diagram of the optical receiver output (back-to-back transmission) with 4-PAM signal, (PRBS $= 2^{11} - 1$, 1 Gbit/s for **a** $P_{opt} = 200\,\mu W$, **b** $P_{opt} = 4\,\mu W$ and **c** $P_{opt} = 20\mu W$) [88]

presented optical receiver. A digital oscilloscope (Lecroy 6100) was used to analyze the received 4-PAM signal.

Eye diagrams of the detected 4-PAM signal are illustrated in Fig. 7.8 at different back-to-back received average optical power values. It is clear from the eye diagrams that the detected 4-PAM signal has equally spaced levels which indicates the good performance and high linearity of the presented receiver with the multilevel signal. In conventional receivers (using limiting amplifiers) this linearity cannot be achieved over a wide optical input power range.

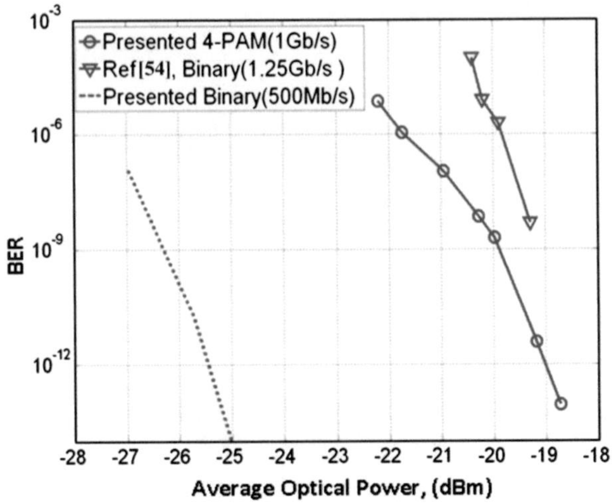

Fig. 7.9 Calculated 4-PAM BER from the measured eye diagram for bit rate 1 Gb/s and binary BER for bit rate 500 Mb/s versus the received back to back average optical power and PRBS $= 2^{11} - 1$ compared to the back-to-back measurements from [54] at 1.25 Gb/s

For sensitivity measurements the 670 nm VCSEL was modulated by a 4-PAM signal with a PRBS length of $2^{11} - 1$ at different received optical powers at the PD. A 1 GHz digital oscilloscope (Lecroy 6100) was used to display the output signal, create the 4-PAM eye diagram and to measure the mean value and the noise variance for each signal level which was used to calculate the symbol error rate SER from relation (2.22).

The calculated BER from the measured eye diagram versus the received back-to-back average optical power for a data rate of 1 Gbit/s and a PRBS $= 2^{11} - 1$ is illustrated in Fig. 7.9 after laser extinction ratio correction from 5 to infinity was made, see Eq. (4.6). The BER for a binary signal at DR $=$ 500 Mbit/s and PRBS $= 2^{11} - 1$ is included in Fig. 7.9.

The proposed optical receiver reaches sensitivity for a 4-PAM signal of -19.8 dBm at a DR of 1 Gbit/s, a wavelength of 670 nm with a BER of 10^{-9} and a PRBS length of $2^{11} - 1$. A sensitivity of -26.3 dBm (BER $= 10^{-9}$) at 500 Mb/s was achieved for a binary signal. There is an improvement by a factor of two in the DR by using a 4-PAM signal, on the other hand there is a measured power penalty of 6.5 dB (theoretical value 4.8 dB). The 1.7 dB extra penalty is due to the laser non linearity and ISI effects.

Back-to-back measurements with a binary signal at a DR of 1.25 Gbit/s with a 658 nm laser diode and PRBS $= 2^7 - 1$ [54] are also included in Fig. 7.9. The receiver consisted of a Si-PIN photodiode (Hamamatsu S9055) with a diameter of 200 μm and 1.5 GHz bandwidth followed by a 900 MHz transimpedance amplifier. The BER was measured with a bit error rate test (BERT) set. Compared to [54], the presented optical receiver shows a better sensitivity, Fig. 7.9. Our receiver nearly has half of the

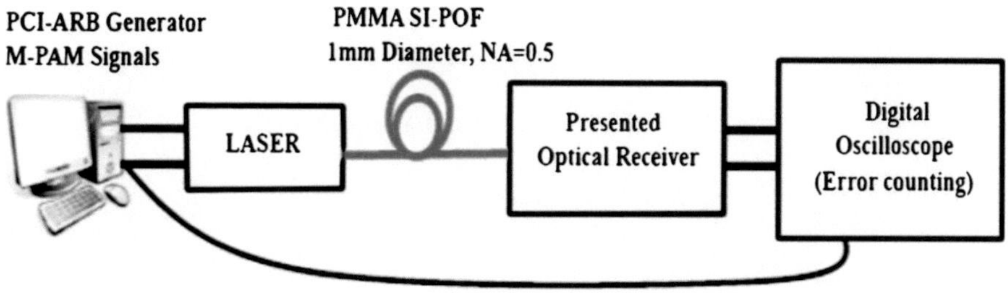

Fig. 7.10 Measurement setup for M-PAM signal transmission over 1 mm diameter PMMA SI-POF with NA = 0.5

receiver bandwidth presented in [54] and nearly the same data rate was transmitted over the two receivers. These results were published in [85, 86].

7.1.3 Gigabit Transmission over SI-POF

A data rate of 1200 Mbit/s over a 10 m long 1 mm-diameter SI-POF(NA = 0.5) was reported in [7] using a 657 nm laser diode with an optical output power of 7 mW and a 640 μm diameter PIN photodiode with a 3 mm diameter optical lens (Hamamatsu-S5052). The BER was measured by BERT and an error-free transmission was achieved for 88 h. To increase the optical receiver bandwidth despite the large-area PD, a 10 Ω broadband optical receiver with 22 V power supply was used. This receiver in [7] consumed several Watts. A pre-distortion filter (high-pass) was added to compensate for the POF small bandwidth. Pre-equalizing of the light source (peaking) by a high-pass filter lowers the light source modulation depth; this will reduce the actual power per pulse compared with rectangular pulses without peaking. It follows that the use of such a method is of particular interest in systems that have adequate power reserves. Also, the used 7 mW optical power laser was beyond eye safety limitations (1 mW for red laser).

To verify our receiver functionality with POF, a 1 Gb/s transmission over 10 m SI-POF was performed [88]. The POF used is the commercially available PMMA SI-POF (POF1.100.B22M) with a core diameter of 1 mm and NA = 0.5. The experimental setup for multilevel transmission over POF is shown in Fig. 7.10. A 1 Gb/s 4-PAM bit stream with 10^8 samples was transmitted over the 10 m long 1 mm-diameter PMMA SI-POF (NA = 0.5). A 670 nm VCSEL (−5 dBm optical power and ER = 5) was modulated by this signal. The received signal from our optical receiver shown in Fig. 7.2 was compared with the original transmitted signal pattern and a real time error counting was performed using a digital oscilloscope (Lecroy 6100) with the built in MATLAB software, Fig. 7.10.

A free of error received signal ($<10^{-8}$) was achieved for a received average optical power of −19 dBm, Fig. 7.11. The number of samples is limited to 10^8 due to the

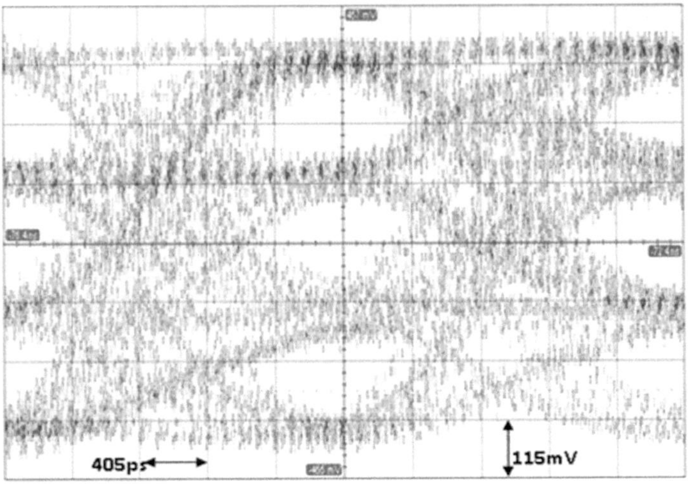

Fig. 7.11 Measured eye diagram of the received optical signal after 10 m length of 1 mm-diameter SI-POF (NA = 0.5) with 4-PAM signal, PRBS = $2^{11} - 1$ at DR = 1 Gbit/s for received average $P_{opt} = -19$ dBm)

limited oscilloscope's memory. Compared to the results in [8], our receiver shows lower power consumption (100 mW) and a lower supply voltage (3.3 V). A laser source within the eye safety optical power range was used [87, 88].

7.1.4 1 Gbit/s over 20 m POF

The 4-PAM signal at a DR of 1 Gbit/s with a PRBS length of $2^7 - 1$ was generated by a multilevel pattern generator. The 4-PAM signal was used to modulate a 655 nm LD with an extinction ratio of ER = 5. The LD is butt coupled to the POF. The optical signal was transmitted through the POF and detected by the integrated optical receiver. The POF is butt coupled to the PD of the receiver shown in Fig. 7.2. A digital oscilloscope (LeCroy 6100) was used to perform BER measurements and to analyze the received 4-PAM signal in real time. The measurement setup is illustrated in Fig. 7.10. The AGC circuit starts to work when the optical input power exceeds −18 dBm. The optical receiver's measured output voltage is constant (750 mV_{PP}). A 1 Gb/s 4-PAM bit stream with 10^8 samples was transmitted over the 20 m long 1 mm-diameter PMMA SI-POF (NA = 0.5) [60]. A 655 nm LD (6 dBm coupled power to the POF) was modulated by 4-PAM signal.
The optical power after the 20 m POF is 3 dBm and the maximum received optical power by the receiver is −5.5 dBm (8.5 dB coupling loss from POF to the integrated photodiode). The received signal from the integrated optical receiver was compared with the original transmitted signal pattern and a real time error counting

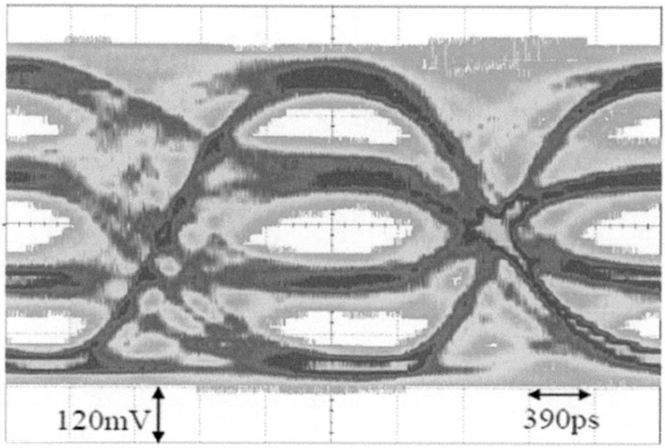

Fig. 7.12 Measured 4-PAM eye diagram of the optical receiver over 20 m PMMA SI-POF with −18 dBm average received optical power at 1 Gbit/s

was performed using a digital oscilloscope (LeCroy 6100) with the built-in MATLAB software. A free of error received signal ($<10^{-8}$) was achieved for a received average optical power of −18 dBm, the eye diagram is illustrated in Fig. 7.12. The number of samples is limited to 10^8due to the limited oscilloscope's memory.
Compared to [5, 7, 33] our system shows a better BER and transmission over longer POF without any equalization and with lower power consumption. Concerning the eye safety, the system is already satisfying the eye safety class 1 M (optical power $\leq$6 dBm) for gigabit transmission over POF [1, 89]. There is an optical power margin of 12.5 dB (free of error signal at −18 dBm). To be more restrictive and to satisfy the eye safety class 1 (LD power is decreased to −4 dBm), a collimating ball lens at the receiver side decreases the coupling loss to 4 dB. Then, the eye safety class 1 can be achieved and there is still an optical power margin of 7 dB.
The design, simulation and measurement results of multilevel transmission over POF using this integrated optical receiver were published in [60, 87, 88].

7.2 Optical Receiver with 200 MHz Bandwidth

7.2.1 Experimental Results

In this section we present another integrated optical receiver with an AGC-TIA and linear post amplifiers. The circuit diagram is like that presented in Fig. 7.2. But a larger diameter PIN-PD was integrated (400 μm) offering the advantage that the positioning of the large diameter POF is easier. The transimpedance feedback resistance (R_{fb}) was increased to have a better sensitivity with a smaller bandwidth of 200 MHz at an optical power of 1 μW and 600 MHz at 500 μW optical power, Fig. 7.13. The

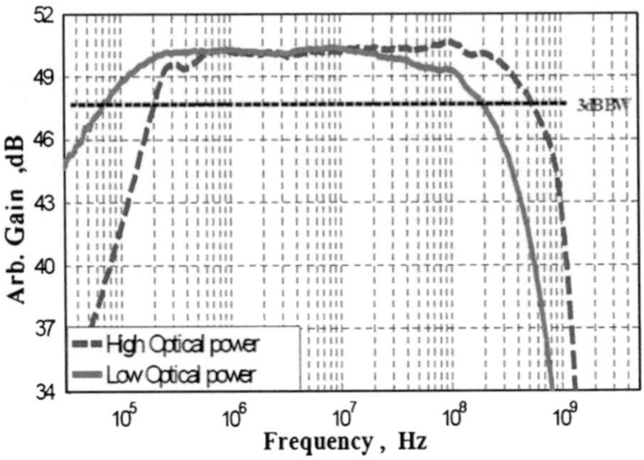

Fig. 7.13 Frequency response of the optical receiver at low input and high input optical power

optical receiver is supplied by a single 3.3 V supply and consumes 90 mW. The input referred noise current for the used TIA is 85 nA. The test chip is fabricated in 0.6 μm BiCMOS technology and has an area of about 1 mm^2 [90].

A fast 670 nm laser diode with an extinction ratio (ER = 4.5) was modulated with a 4-PAM signal with a PRBS length of $2^9 - 1$. The AGC circuit starts to work at an optical input power of −23 dBm and the measured output voltage is constant with 380 mV_{PP} over the input range from −23 dBm to the maximum available laser source power (−5 dBm). The simulation results indicate that the optical receiver's input dynamic range extends from −23 to 0 dBm. The measured 4-PAM eye diagram shows equally spaced levels with constant amplitude, see Fig. 7.14. This equally spaced output voltage levels with fixed amplitude make the detection and decoding of a 4-PAM signal easy at fixed threshold levels. The equally spaced levels within a wide input power range prove the optical receiver's high linearity in receiving multilevel signal. This is very difficult with conventional optical receivers having limiting amplifiers or small linear range AGC amplifiers.

A receiver published in [91] with an AGC-TIA and two integrated 430 μm diameter photodiodes, using a binary signal at a DR of 150 Mbit/s had an estimated sensitivity of −28 dBm at 10^{-9} BER (post-layout simulation), whereas our receiver presented here shows a measured sensitivity of −29.5 dBm (1.5 dB improvement) at 200 Mbit/s with binary signal. For 400 Mbit/s the presented receiver shows −27 dBm sensitivity at BER = 10^{-9}.

For sensitivity measurements, a fast 670 nm laser diode was modulated with a 4-PAM signal with a PRBS length of $2^9 - 1$ at different optical power. A laser ER correction from 4.5 to 10 was made. A Lecroy 6100 oscilloscope was used to analyze the output signal eye diagram, to determine the mean voltage level and the noise variance for each level, which was used to calculate the symbol error rate (SER).

The measured BER of the presented receiver versus average optical input power for 4-PAM modulation at a DR of 400 Mbit/s is illustrated in Fig. 7.15 [90]. Results of

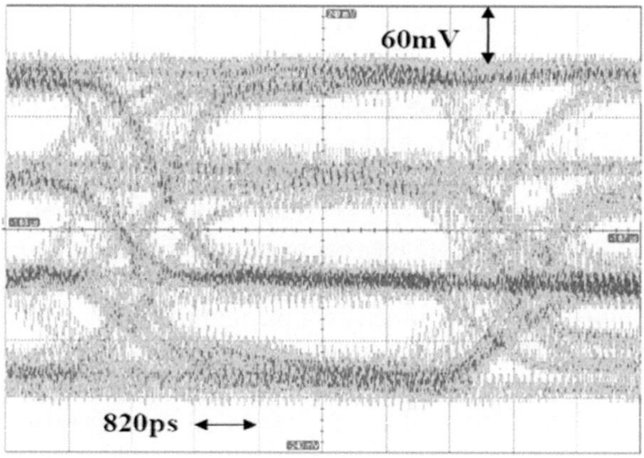

Fig. 7.14 Measured eye diagram of the optical receiver 4-PAM signal (PRBS $= 2^9 - 1$, 500 Mbit/s, $P_{opt} = -20$ dBm and ER = 4.5)

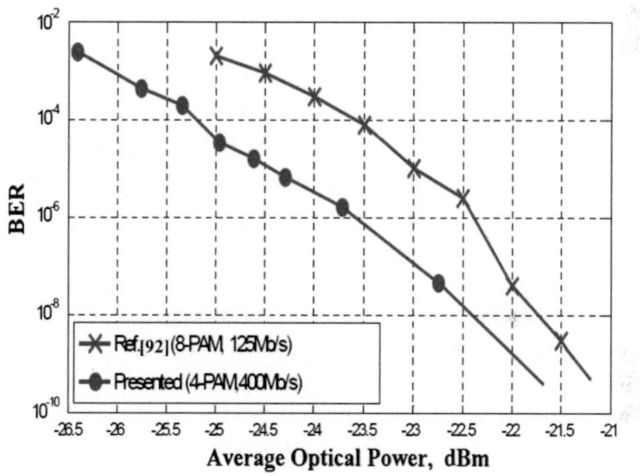

Fig. 7.15 Measured BER versus average optical power for the presented optical receiver and the receiver used in [92]

back-to-back transmission using an 8-PAM signal working at a DR of 125 Mbit/s to modulate a 520 nm LED [92] are also included in Fig. 7.15. Reference [92] used an optical receiver consisting of an integrated PD with a transimpedance amplifier (Hamamatsu S6468-02) and an automatic gain control AGC. Compared to [92], our presented optical receiver has an about 1 dB sensitivity improvement at a more than three times higher data rate of 400 Mb/s. The high linearity of the presented AGC optical receiver with multilevel signals over a wide input power range is the reason for its better performance than other published AGC receivers. The integration of the PD increases the receiver's sensitivity and bandwidth.

7.2.2 500 Mbit/s Transmission over SI-POF

Test signal is a 500 Mb/s (baud rate 250 Mbaud) 4-PAM and $2^{11}-1$ PRBS bit stream generated by a PCI arbitrary waveform generator (ARB). The ARB was programmed to generate the 4-PAM signal. This drives a 670 nm laser source. A single-mode fiber pigtail is attached to the laser diode. This single-mode fiber pigtail is not necessary for the system to work, but makes it easier to set up different experiments. In a low-cost system, the same laser diode (without fiber pigtail) can be used in combination with a small collimating lens to directly couple into the POF.

From the single-mode fiber, the laser beam is launched into 10 m of commercially available PMMA SI-POF (POF1.100.B22M) with a core diameter of 1 mm and NA = 0.5. At the output of the POF a collimating lens used to focus the laser beam on the PIN photodiode (integrated with the presented receiver) with a diameter of 400 μm. This large-area PD enables to collect and detect most of the transmitted optical signal. A 1 GHz digital oscilloscope (Lecroy 6100) was used to display the output signal. The signal was shown as a 4-PAM eye diagram and the mean value and the noise variance for each signal level was measured. These were used to calculate the symbol error rate (SER) for a 4-PAM signal [93].

The transmitted average optical power from the laser source was −5.5 dBm. The coupling between the glass optical fiber and the POF caused a 4 dB loss. The PMMA POF has an attenuation of 0.3 dB/m at the used 670 nm wavelength; this will result in 3 dB attenuation over 10 m. To collect most of the transmitted signal the position of the receiver to the collimating lens was varied until the average PD monitoring current reached the maximum. The collected average optical power was −19.5 dBm at 500 Mb/s. This indicates that there is a 7 dB loss due to coupling from POF to the PD via the collimating lens. This loss can be decreased in the future by improving the lens system. Figure 7.16 shows the measured eye diagram of the signal after transmission over 10 m PMMA SI-POF with 1 mm diameter at a DR of 500 Mb/s and PRBS length $2^{11}-1$. The received optical power after 10 m SI-POF is −19.5 dBm.

The peak-to-peak optical receiver output voltage is fixed and equal to 380 mV over a wide range of the input optical power. The eye diagram for the 4-PAM signal shows error-free transmission ($Q > 6$) and equally spaced levels. This equally spaced signal makes the decoding of the 4-PAM signal to a binary stream very easy compared to the unequally spaced levels which will result from simple limiting (or switched) amplifiers.

By using the measured mean value and the standard deviation for each level, the BER for each eye can be calculated. The BER for the first eye is $5\cdot10^{-13}$, the second eye's BER is $5\cdot10^{-14}$ and the third eye's BER is $2\cdot10^{-12}$. The measured SER is $4\cdot10^{-12}$ which is nearly equal to the worst BER of the three eyes. The coupling efficiency can be increased by using an aspheric lens between POF and receiver. There, the coupling loss can be decreased from 7 to 3 dB. Also, a 650 nm laser source will reduce the fiber attenuation to 0.15 dB/m compared to 0.3 dB/m at 670 nm. The integrated optical receiver design and the measured results for multilevel transmission over SI-POF were published in [90, 93, 94].

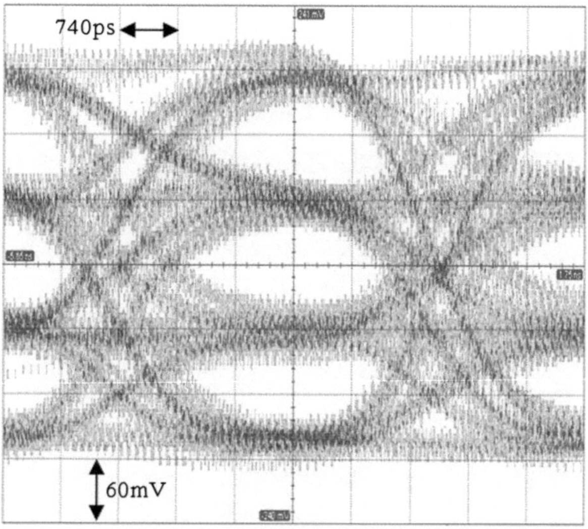

Fig. 7.16 Measured eye diagram for 500 Mbit/s 4-PAM signal with PRBS length $= 2^{11} - 1$ after 10 m PMMA SI-POF for the received average optical power of −19.5 dBm

To reach 500 Mbit/s over 30 m PMMA SI-POF a higher optical power laser source is needed. A 500 Mb/s 4-PAM bit stream with 10^8 samples was transmitted over the 30 m long 1 mm-diameter PMMA SI-POF (NA= 0.5). A 655 nm LD (6 dBm coupled power to the POF) was modulated by a 4-PAM signal. The optical power after the 30 m POF is 2.3 dBm and the maximum received optical power by the optical receiver is −5 dBm (7.3 dB coupling loss from the 1 mm POF to the integrated 400 μm PD). The received signal from the integrated optical receiver was compared with the original transmitted signal pattern and a real time error counting was performed using an oscilloscope LeCroy 6100 with the built-in MATLAB software. A free-of-error received signal ($<10^{-8}$) was achieved for a received average optical power of −20 dBm, the 4-PAM eye diagram is illustrated in Fig. 7.17.

7.3 High Linearity Optical Receiver with 112 MHz Bandwidth

Many researchers prove that by using multicarrier modulation schemes like OFDM the POF bandwidth limitations can be overcome. OFDM is seen as promising technology for low-cost Gigabit transmission over SI-POF [6].

A major problem of OFDM is that the multicarrier signals usually have a large envelop fluctuation. This means that the peak-to-average power ratio (PAPR) of multicarrier signals can be high. These large peaks increase the amount of intermodulation distortion (IMD) resulting in an increase in the error rate.

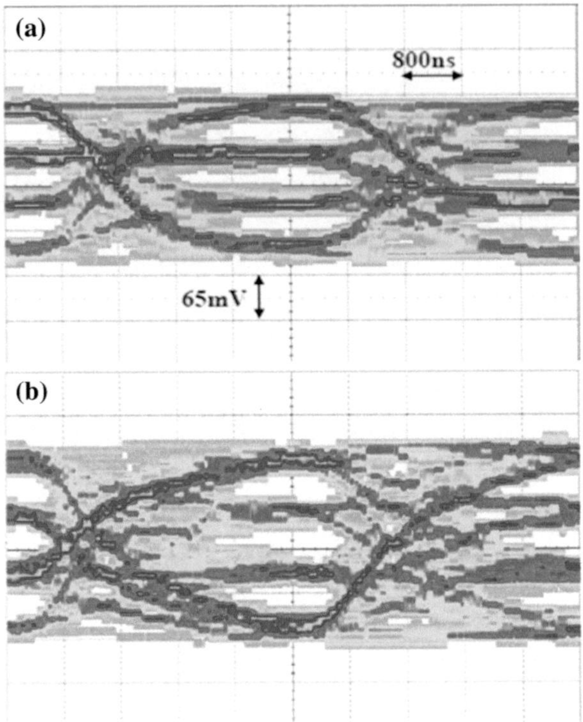

Fig. 7.17 Measured eye diagram for 500 Mbit/s 4-PAM signal with PRBS length = $2^7 - 1$ after 30 m PMMA SI-POF for the received average optical power **a** −20 dBm and **b** −4.5 dBm

A high-linearity receiver with AGC is required for OFDM to reduce the IMD. The AGC adjusts the gain of the amplifier to obtain the desired output signal strength. In conventional AGCs the signal amplitude is typically estimated using a peak detector. This adequately works for signal amplitudes with fixed PAPR such as sinusoids, but is unreliable for high-PAPR signals. These issues excluded the use of a conventional AGC with OFDM [95]. The wireless local area network receiver presented in [95] implemented an AGC, which uses RMS detection. This was found to be adequate for estimating the OFDM amplitude by integrating for the duration of a training symbol. In this section we present an integrated linear optical receiver consisting of a TIA with AGC followed by an amplifier and a line driver. Additionally two main control loops are present, one for background-light (BGL) cancellation and one for the gain control. The speciality of this linear optical receiver is that the gain is only controlled by BGL, which is always present in optical transmission systems due to laser diode biasing above the threshold (limited extinction ratio). Compared to peak-detecting AGC commonly used a better linearity can be achieved with our BGL-AGC especially for pulsed signals or more general signals with a high PAPR where the pulse repetition rate is low. When peak detection is used with this kind of signals,

large storage capacitors would be necessary to store the peak value. Additionally large capacitors cannot easily be charged very fast and accurate. We do not need these large capacitors with the BGL-AGC [61, 96, 97].
Nearly all the previous attempts to design a linear optical receiver for short-range communication include only simulation results. No measurements were presented to verify the optical receiver linearity. A CMOS common-gate differential TIA was presented in [98] to obtain a better linearity and to reduce the IMD. A 40 dBc third-order intermodulation product (IMP3) was obtained using HSPICE. Reference [99] has shown analytically that there are significant benefits in linearity when choosing common-emitter feedback (CEFB) over traditional resistive feedback. There, CEFB had an output third-order intercept point of 23 dBm. A TIA which divides the feedback resistor into shunt and series-connection was introduced in [100]. The eye diagram simulations showed a better linearity for the new TIA topology compared to the conventional topology.

7.3.1 Circuit Description

Figure 7.18 shows the circuitry of the integrated optical receiver. There are two identical 400 μm diameter Si PIN-PDs. One receives the optical signal and the other is kept dark. The integrated large-area photodiode (PD) makes it easy to couple and position the large-core POF.
The dark PD is used to balance the differential TIA which has two amplifying stages and two feedback resistors (R_{fb}). The differential topology of the TIA has the advantage of high immunity against common-mode and power supply noise. The input referred noise current of this TIA is 70 nA. The PD antireflection coating is optimized for red light with a responsivity of 0.5 A/W.
The first stage is a differential MOSFET amplifier (M_3, M_4) which is shunted by PMOSFETs (M_1, M_2) to reduce the gain of the TIA by a steering voltage. The second stage of the TIA is a linear amplifier (LA1) consisting of a bipolar differential stage with emitter followers. This configuration of the TIA has two big advantages: first a very wide dynamic range of the TIA can be achieved without any stability problems, therefore complex and performance critical capacitor switching can be avoided. The second advantage is that large photocurrents (I_{ph}) in the mA-range can be processed if M_1 and M_2 are fully conductive, since the drain current of the first stage is also in the order of mA to achieve overall low noise.
The first control loop cancels the BGL by injecting a current via M_7 into the dark side of the TIA. M_7 is steered by an integrator (I1), which senses the differential voltage of the TIA. Therefore the average differential voltage at the TIA output is held at zero. M_5 also generates photocurrent (I_{ph}); therefore it is connected to a dummy structure consisting of R_{fb2} and M_{10}. Basically the dummy defines the current transimpedance of the TIA; therefore with a second control loop (I2) the AGC-voltage is controlled in such a way that the voltage at the dummy is held constant for all possible I_{ph}. This voltage is defined by V_{ref2}, and it is a representation of the maximum linear range

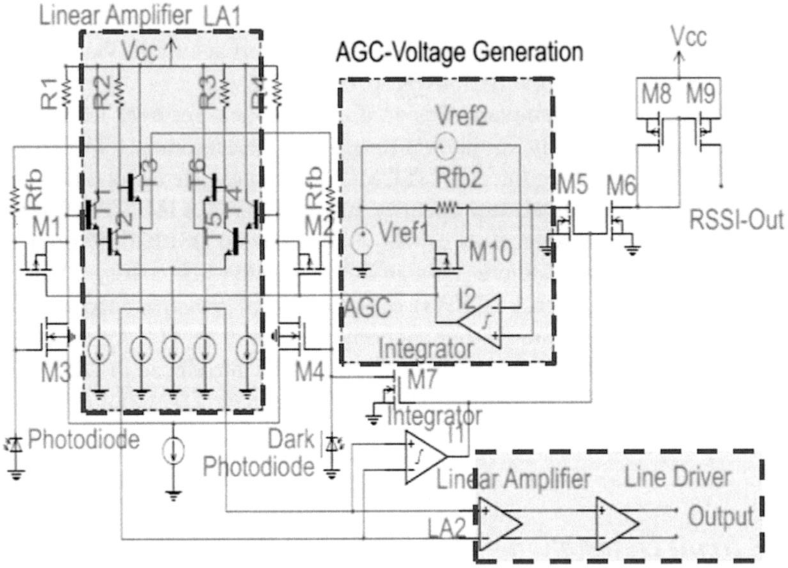

Fig. 7.18 High linearity integrated optical receiver

of the TIA. V_{ref1} is a reference voltage of the dummy circuit of R_{fb2} and M_{10} which give a bias voltage equal to the voltage at the TIA (M_3 and M_4 drain voltages). The output voltage will be constant over a wide input optical power range for a fixed extinction ratio (ER). M_6, M_8 and M_9 replicate the average I_{ph} to a received signal strength indication (RSSI) output for diagnostic purposes.

An AGC sets the gain of the TIA to avoid overloading of the TIA and to increase the dynamic range of the optical receiver in the linear region. This wide linear input dynamic range makes the receiver ideally suited for OFDM signals.

The AGC circuitry decreases the TIA gain for increasing total input optical power to ensure that the output voltage of the TIA is constant over a wide input dynamic range. As a result, each level of the output multilevel signal will be amplified by the same gain, so the output signal will have equally spaced voltage levels. The AGC outputs control the gate voltage of the PMOS transistors M_1 and M_2. As the average input optical power increases the AGC output voltage decreases. This leads to a decrease in the PMOS resistance of M_1 and M_2. Decreasing the PMOS resistance has the effect of decreasing the gain of the TIA. This is due to the shunt configuration between the PMOS and the feedback resistor R_{fb}.

The output signal of the TIA is amplified by a linear post amplifier LA2. The last stage in the receiver, which forms the interface to the following off-chip signal processing unit, is an emitter follower which can drive a load current up to 24 mA.

The integrated optical receiver is supplied by a single 3.3 V supply and consumes 33 mW without output load. The test chip is fabricated in 0.6 μm technology and has an area of about 0.87 mm^2 (0.84 × 1.04 mm). The optical receiver chip micrograph is shown in Fig. 7.19.

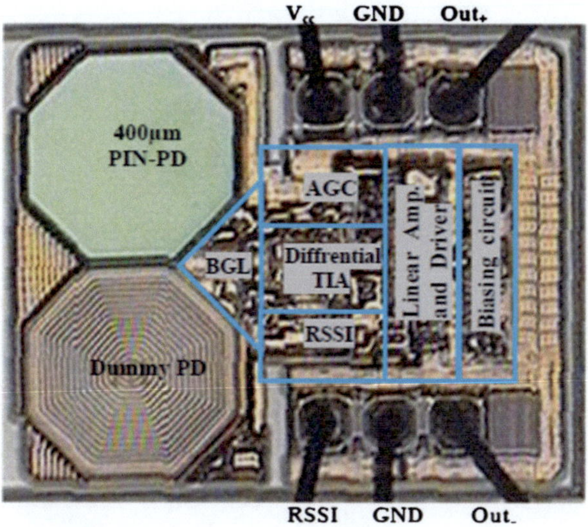

Fig. 7.19 Die photo of the high linearity optical receiver

7.3.2 Experimental Results

The optical receiver's frequency response was measured with a network analyzer (HP 8753E). A calibration was performed to cancel the effect of the laser source and the connected measurement cables on the receiver frequency response.
The measured receiver's −3 dB bandwidth is 112 MHz at maximum amplification (300 kΩ) and 280 MHz at the lowest amplification (2.5 kΩ). With the 112 MHz bandwidth a data rate of 250 Mb/s can be received using a binary signal. The data rate can be doubled to 500 Mb/s by using 4-PAM signaling at the same bandwidth [61, 96].
For sensitivity measurements, a binary and a 4-PAM signal with a symbol rate of 250 Mbaud were generated on a personal computer with MATLAB software. The signals were transmitted by a 12 bit arbitrary waveform generator PCI-Card (Acquitek DA11000-16M). The generated signals were used to modulate a commercial low price 655 nm laser diode with an extinction ratio of ER = 2.5. The optical signal from the laser was transmitted through the single-mode glass optical fiber (GOF) and detected by the presented optical receiver. The sensitivities of the optical receiver for binary and 4-PAM signals were measured at different average optical power values. The measured four-level eye diagram shows the receiver's high linearity when the received average optical power changes from 5 to 650 µW (maximum available power in our lab); equally spaced voltage levels and a constant output swing are illustrated in Fig. 7.20. The post-layout simulation shows that the receiver can work with a high linearity up to 800 µW average input optical power.
A 1 GHz digital oscilloscope (Lecroy 6100) was used to display the output signal, create the 4-PAM eye diagram and to measure the mean value and the noise variance

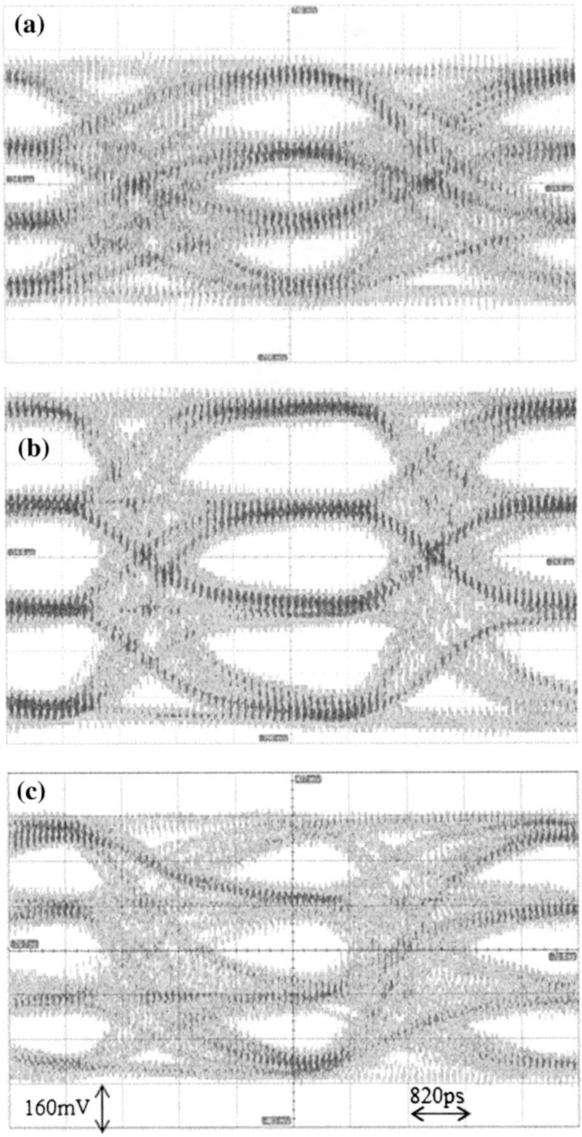

Fig. 7.20 Measured 4-PAM eye diagrams of the optical receiver with back-to-back transmission (1 m GOF) at PRBS = $2^{11} - 1$, DR = 500 Mb/s, ER = 2.5, and received optical power = 8 μW **a**, **b** received optical power = 650 μW, **c** after 40 m PMMA SI-POF with received optical power = 9 μW

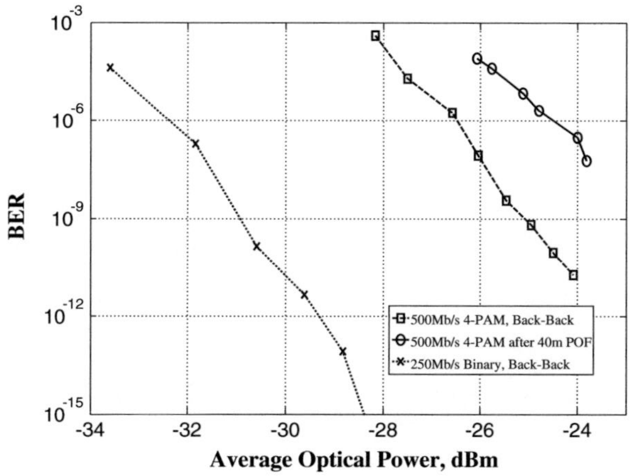

Fig. 7.21 Measured BER versus received average optical power for binary and 4-PAM signals with back-to-back transmission (1 m GOF) as well as after 40 m SI-POF

for each signal level which was used to calculate the SER for 4-PAM signal, relation (2.22).

The measured BER versus the received average optical power for DR = 500 Mbit/s and PRBS $= 2^{11} - 1$ is illustrated in Fig. 7.21. The BER for a binary signal at DR = 250 Mbit/s and PRBS $= 2^{11} - 1$ is also included in Fig. 7.21. The results in Fig. 7.21 consider a laser extinction ratio (ER) correction from 2.5 to infinity. There is a power penalty of 6 dB for 4-PAM compared to binary (4.8 dB theoretical). The extra 1.2 dB penalty is due to laser nonlinearity and ISI effects.

The proposed optical receiver reaches sensitivity for binary signal of −31 dBm at a DR of 250 Mbit/s at a wavelength of 655 nm with a BER of 10^{-9} and a PRBS length of $2^{11} - 1$. Another receiver presented in [91] with AGC-TIA and two integrated 430 μm diameter PDs had an estimated sensitivity of −28 dBm at 10^{-9} BER using a binary signal at a DR of 150 Mbit/s [91]. Our presented receiver shows a better measured sensitivity (3 dB improvement) at nearly the double data rate.

The presented optical receiver shows a sensitivity of −25 dBm (BER $= 10^{-9}$) at DR = 500 Mb/s using a 4-PAM signal. There is an improvement by a factor of two in the DR by using the 4-PAM signal compared to binary signal. On the other hand there is a measured power penalty of 6 dB (theoretical value 4.8 dB for M = 4).

For linearity measurements, the 675 nm laser source with an ER = 3 was modulated by a single tone (20 MHz). The measured THD was 1 % at P_{Opt} equal to −23 dBm and 1.7 % at −3.5 dBm, Fig. 7.22. The receiver THD is actually better, since the laser source alone has a THD of 0.7 %. The receiver's simulated THD is less than 0.5 and 0.3 % for P_{Opt} equal to −3.5 and −23 dBm, respectively [97].

The IMP3 were measured at different P_{Opt} values. The fundamental measured power (two tones at 100 and 101 MHz) and IMP3 are plotted in Fig. 7.23. The AGC starts to work at a P_{Opt} of −23 dBm. The fundamental frequency of the output signal has

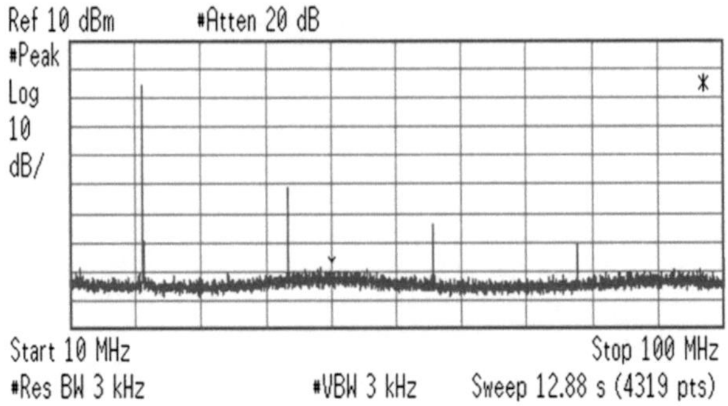

Fig. 7.22 Measured harmonics spectrum for the optical receiver with single-tone input (20 MHz) at −3.5 dBm input average optical power

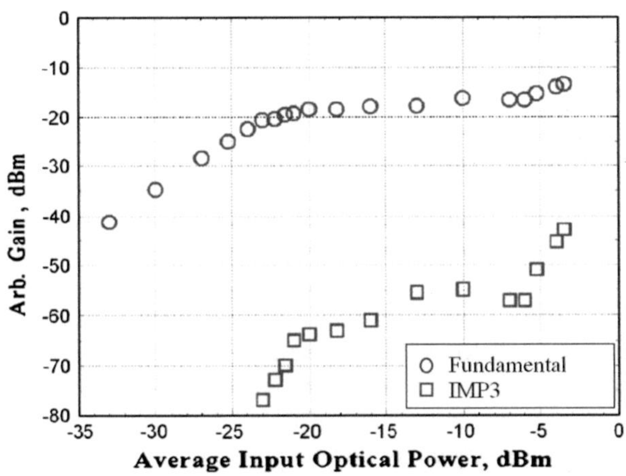

Fig. 7.23 Measured two-tone third-order intermodulation (IMP3) and fundamental power for fundamentals of 100 and 101 MHz at different input average optical power values

a constant output power when P_{Opt} increases from −23 to −6 dBm. The IMP3 is less than −40 dBc at P_{Opt}less than −6 dBm. For lower frequencies (two tones at 20 and 21 MHz) the optical receiver shows an IMP3 of −50 dBc at $P_{Opt} = -3.5$ dBm (Fig. 7.24).

For comparison, the AGC introduced in [101] fabricated in 0.25 μm SiGe BiCMOS technology with an active area of 0.225 mm^2 has a THD of 0.9 % at the lowest gain and its low and high cut off frequencies are 68 kHz and 95 MHz, respectively. The presented BGL-AGC circuit has a chip area of 0.16 mm^2, measured cut off frequencies of 0.6 kHz and 112 MHz, respectively, and a THD of 0.5 % at the lowest gain. This comparison shows that our AGC circuit has a better linearity, wider bandwidth and needs a smaller chip area. Also the convergence of the BGL-AGC was simulated using SPECTRE with an OFDM signal.

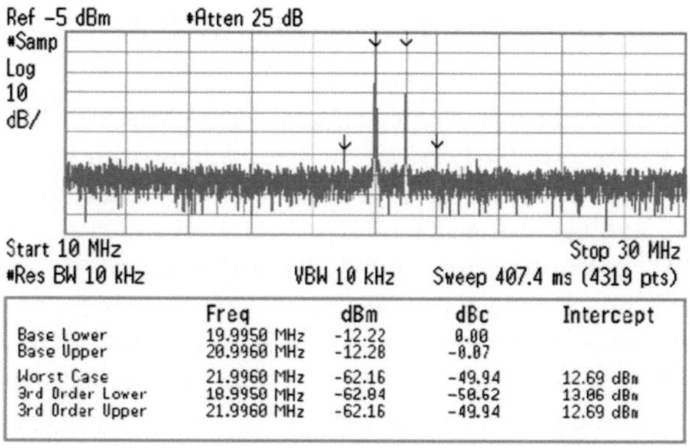

Fig. 7.24 Two-tone third-order intermodulation for fundamentals of 20 and 21 MHz at −3.5 dBm input average optical power

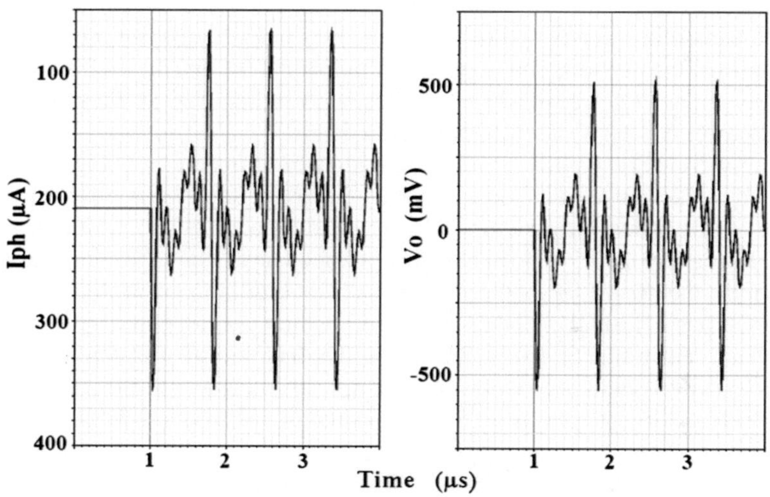

Fig. 7.25 AGC performance with OFDM signals in low-gain adjustment for a BGL optical input power of −3.5 dBm when modulation starts at 1 μs (*left*: input photocurrent, *right*: receiver output)

The result for low-gain adjustment from a high input P_{Opt}, is shown in Fig. 7.25. At 1 μs the modulation starts and the output signal is constant from onset of modulation. The BGL is always present and constant in a certain optical system, so the BGL-AGC adjusts signals to the desired level instantaneously. As a conclusion the specialty of the presented linear optical receiver is that the gain is only controlled by the background-light (BGL) photocurrent, which results in immediately constant output signals. Compared to peak-detecting AGC commonly used, a better linearity can be achieved with BGL-AGC especially for signals with a high PAPR.

7.3.3 500 Mbit/s Transmission over 40 m SI-POF

The sensitivity measurement was repeated with 40 m PMMA SI-POF replacing the 1 m GOF link. The received 4-PAM signal eye diagram with 9 μW received optical power at DR = 500 Mb/s over 40 m POF is illustrated in Fig. 7.20c. The eyes are more closed than the eyes in Fig. 7.20a (back-to-back) due to the intersymbol interference (ISI) coming from the limited SI-POF bandwidth. The SER calculation using Eq. (2.22) is now invalid since the channel shows ISI [61, 96].
A 4-PAM signal with 10^8 samples was transmitted over the 40 m POF and the received signal from the optical receiver was compared with the original transmitted signal. Real-time error counting was performed using an oscilloscope (Lecroy 6100) with built-in MATLAB. The measured 4-PAM BER versus average optical power for DR = 500 Mbit/s and PRBS = $2^{11} - 1$ over 40 m POF is included in Fig. 7.21. A received signal free-of-error (BER $< 10^{-8}$) was achieved for an average optical power larger than −23 dBm. The number of samples is limited to 10^8 due to the limited oscilloscope memory. There is a 2 dB difference between the BER with back-to-back transmission and transmission over 40 m POF because of the limited bandwidth of the SI-POF. The frequency responses of the optical receiver with 40 m POF and 0.5 m POF transmission are illustrated in Fig. 7.26 at the same average optical power of 12 μW at a wavelength of 655 nm.
The received electrical power measured as function of frequency in Fig. 7.26 is proportional to the square of the optical power, due to the square-law effect of the PD. So, the −3 dB optical bandwidth will be the −6 dB electrical bandwidth. Increasing the POF length from 0.5 to 40 m has a big effect on decreasing the complete system bandwidth; the 40 m POF is the main frequency limiting part in the system. The 40 m POF has a 110 MHz optical bandwidth which can support a symbol rate of 250 Mbaud. For a binary signal, increasing the data rate by a factor of two (from 250 to 500 Mbit/s) will need twice the POF optical bandwidth (220 MHz) where the power in Fig. 7.26 dropped by −15 dB compared to −6 dB at 110 MHz (9 dB power penalty).
By using 4-PAM signaling a DR of 500 Mb/s is reached over the 110 MHz POF bandwidth and this 9 dB power penalty is saved compared to 500 Mb/s binary streams. Theoretically, the frequency response of multimode POF shows a Gaussian-like behavior [6, 49].

$$P(f) = P_o exp^{-(f^2/f_o^2)} \tag{7.1}$$

where P_O is the output signal power at low frequency, f is the output signal frequency, $f_o = f_{3dB}/\sqrt{\ln(2)}$ and f_{3dB} is the −3 dB bandwidth of the full electrical-optical electrical channel. Therefore, by shrinking the transmission bandwidth to the half, a level margin of 15 dB or more can be obtained for longer POF length.

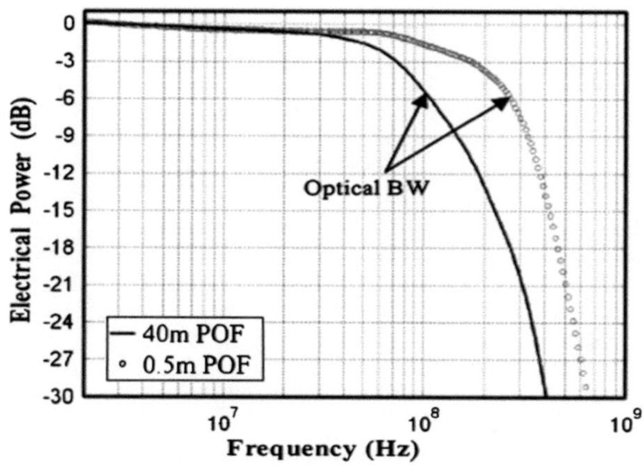

Fig. 7.26 Measured frequency response for the 655 nm laser source and optical receiver with 0.5 and 40 m POF

7.3.4 400 Mbit/s Transmission over 50 m POF

Another experiment was performed with an 8-PAM signal to extend the POF transmission length to 50 m. The received signal from the optical receiver was compared with the original transmitted signal. This real-time error counting was performed using a digital oscilloscope (Lecroy 6100) with built-in MATLAB. A received signal free-of-error (BER $< 10^{-8}$) was achieved for an average optical power larger than -19 dBm. The number of samples is limited to 10^8 due to the limited oscilloscope memory [62].

The received electrical power measured as function of frequency in Fig. 7.27 is proportional to the square of the optical power, due to the square-law effect of the PD. So, the -3 dB optical bandwidth will be the -6 dB electrical bandwidth. The 50 m POF has a 93 MHz optical bandwidth which can support a symbol rate of (93/0.7) Mbaud = 132 Mbaud. For a binary signal, increasing the data rate by a factor of three (from 132 to 400 Mbit/s) will need three times the POF optical bandwidth (280 MHz) where the power in Fig. 7.27 dropped by -18 dB compared to -6 dB at 93 MHz (12 dB power penalty).

By using 8-PAM signaling, a DR of 400 Mb/s is reached over the 93 MHz POF bandwidth and this 12 dB power penalty is saved compared to 400 Mb/s binary streams. The measured eight-level eye diagram for 400 Mb/s over 50 m POF with 20 μW received optical power shows the receiver's high linearity; equally spaced voltage levels are illustrated in Fig. 7.28.

According to the literature a lot of previous efforts were made to transmit more than 400 Mbit/s over 1 mm core diameter PMMA SI-POF with NA = 0.5. An error-free transmission with a data rate of 400 Mbit/s over a 10 m long PMMA SI-POF was achieved. The binary data rate extended to 622 Mbit/s over only 1 m POF with 10^{-11} BER [102, 103]. The light source was a 650 nm RC-LED. The used receiver

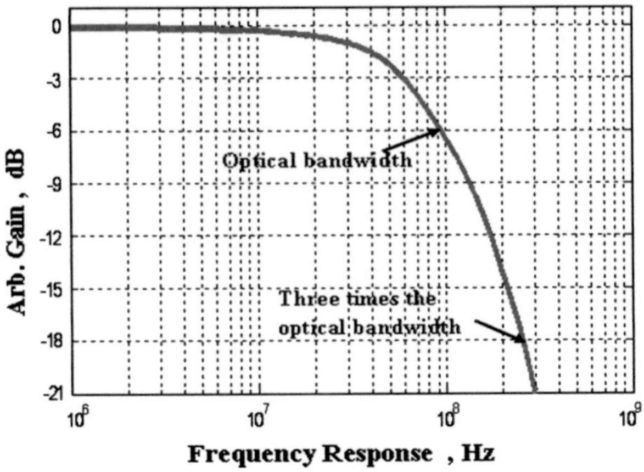

Fig. 7.27 Measured frequency response for the optical receiver after 50 m PMMA SI-POF

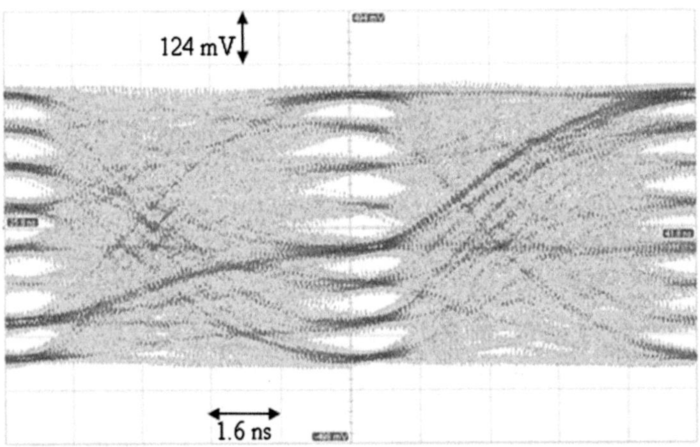

Fig. 7.28 Measured 8-PAM eye diagram of the optical receiver over 50 m PMMA SI-POF with 20 μW received optical power at DR = 400 Mb/s, PRBS = $2^{11} - 1$ and an extinction ratio of 3

was simply a Hamamatsu PIN photodiode. Using only a photodiode as an optical receiver causes a lower sensitivity (more optical power is needed) than with PD and TIA.

A 500 Mbit/s transmission over 50 m POF with 650 nm RC-LED was possible using PAM-4 signaling and equalization [9]. The signal is received by a discrete PIN photodiode with an active area of 0.8 mm^2 and an amplifier. For the case without using equalization the eye diagram is completely closed [9], so an error-free transmission is not possible. A feed-forward equalizer (FFE) with 5 taps was needed to have an error-free signal (offline processing). The dispersion compensation in [9] behind the

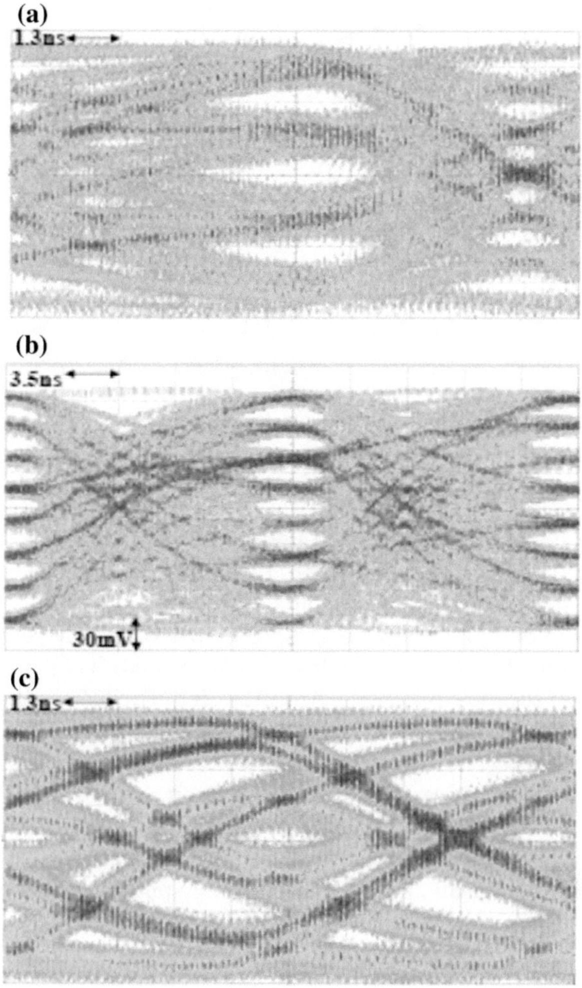

Fig. 7.29 170 Mb/s over 115 m SI-POF using different modulation formats **a** 4-PAM **b** 8-PAM **c** binary, at −22 dBm average received optical power

receiver has a bad effect of amplifying the noise at higher frequencies. Our optical receiver shows an error-free signal ($<10^{-8}$) over 50 m PMMA SI-POF without using any equalization.

1 Gbit/s transmission is experimentally demonstrated over a 50 m SI-POF [104], with a 650 nm RC-LED at a BER of $4 \cdot 10^{-4}$. This is achieved by adding a pre-equalizing (high-pass filter) and DMT. The front-end optical receiver used was a discrete 0.8 mm-diameter Si photodiode with separate transimpedance amplifier. The use of DMT modulation dramatically increases the receiver complexity and cost. ADC and FFT processor chips are required which increase the power consumption

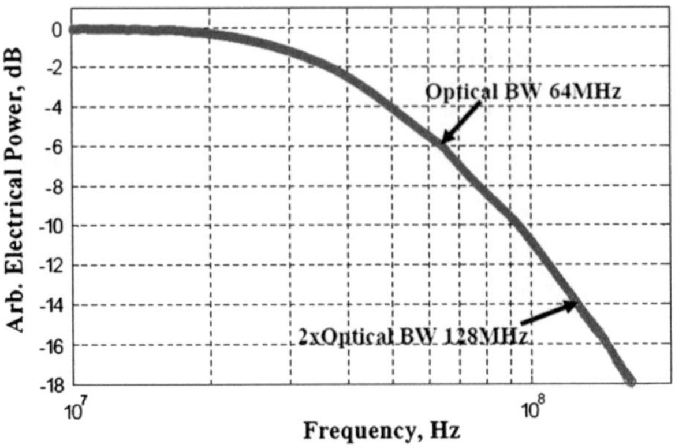

Fig. 7.30 Measured frequency response of the high linearity optical receiver after 115 m PMMA SI-POF

and the system costs. Pre-equalizing of light source (peaking) by a high-pass filter lowers the light source modulation depth; this reduced the actual power per pulse compared with rectangular pulses without peaking.

All of these measures of post- or pre-equalization used in [9, 104] are at the expense of the power budget. It follows that the use of such methods is of particular interest in systems that have adequate power reserves. Also, if the frequency response changes, as a result of different lengths of the POF or a bend in the fiber, the result will be too much or too little compensation so that the pulses become distorted.

The presented optical receiver shows an error-free signal over 50 m PMMA SI-POF without using any equalization which has the advantage of saving the system power budget of pre-equalization. The integrated optical receiver design and measured results of multilevel transmission over SI-POF were published in [61, 62, 96, 97].

7.3.5 170 Mbit/s Transmission over 115 m SI-POF

Another experiment was performed with binary, 4-PAM and 8-PAM signals to extend the POF transmission length to 115 m [68]. The received signal from the integrated optical receiver introduced in Fig. 7.18 was compared with the original transmitted signal. This real-time error counting was performed using a digital oscilloscope (Lecroy 6100) with built-in MATLAB. By transmitting 170 Mbit/s by 4-PAM and 8-PAM a free of error (BER $\ll 10^{-8}$) received signal was achieved for a received average optical power of -22 dBm. The measured 4-PAM and 8-PAM eye diagrams for 170 Mb/s over 115 m POF with -22 dBm are illustrated in Fig. 7.29a, b. For binary transmission at 170 Mbit/s over 115 m POF, the error rate was very high and the eye diagram was completely closed, see Fig. 7.29c.

The received electrical power measured as function of frequency in Fig. 7.30 is proportional to the square of the optical power, due to the square-law effect of the PD. So, the −3 dB optical bandwidth will be the −6 dB electrical bandwidth. The 115 m POF has a 64 MHz optical bandwidth which can support a symbol rate of 85 Mbaud. For a binary signal, increasing the data rate by a factor of two (from 85 to 170 Mbit/s) will need two times the POF optical bandwidth (128 MHz) where the power in Fig. 7.30 dropped by −14 dB compared to −6 dB at 64 MHz (8 dB power penalty). By using 4-PAM signaling a DR of 170 Mb/s is reached over the 64 MHz POF bandwidth and this 8dB power penalty is saved compared to 170 Mb/s binary streams. The net improvement is 8 dB − 5.1 dB = 2.9 dB, see relation (2.13).

Chapter 8
Equalizer Implementations for SI-POF

In this chapter we introduce optical receivers with POF equalizers to enhance the performance for giga-bit communication over SI-POF. A single-chip fully integrated optical receiver with an integrated POF-equalizer will be introduced. The integration of large-area photodiodes, TIA, POF equalizer and 50 Ω driver on a single chip enhances the performance and lowers the cost for giga-bit communication over SI-POF. The different types of SI-POF equalizers for giga-bit transmission will be discussed. The implementation, experimental results, and discussions of these POF equalizers will be presented.

8.1 Equalized Giga-Bit Transmission over SI-POF

There are two approaches for a SI-POF equalizer in the literature. The first is to use a pre-equalization and the second is to use post-equalizers.

Reference [105] used passive pre-equalization to reach a data rate of 1.25 Gbit/s over 50 m SI-POF for binary NRZ modulation. A 650 nm edge emitting laser diode, an optical receiver with an 800 μm diameter PIN-PD and a separate TIA were used. The measured sensitivity was −16 dBm at a BER of 10^{-9}.

Laser source pre-equalization (peaking) lowers the modulation depth of the emitted light; this reduces the effective power per pulse compared with rectangular pulses without peaking. This is at the expense of the system power budget.

Also, if the frequency response changes, as a result of different lengths of the POF or bends in the fiber, the result will be too much or too little compensation, therefore the BER will increase.

In [65] a data rate of 1 Gb/s over 50 m POF was achieved by using an adaptive Decision Feedback Equalizer (DFE) and Forward Error Correction (FEC) which were fully implemented in an FPGA. The measured sensitivities for BER= 10^{-9} were −13.5 dBm using a resonant cavity light emitting diode (RC-LED) and −18.5 dBm with a VCSEL. The adaptive post-equalization can overcome the changes of the frequency response resulting from different POF lengths or POF bending.

M. Atef and H. Zimmermann, *Optical Communication over Plastic Optical Fibers*, Springer Series in Optical Sciences 172, DOI: 10.1007/978-3-642-30388-3_8,

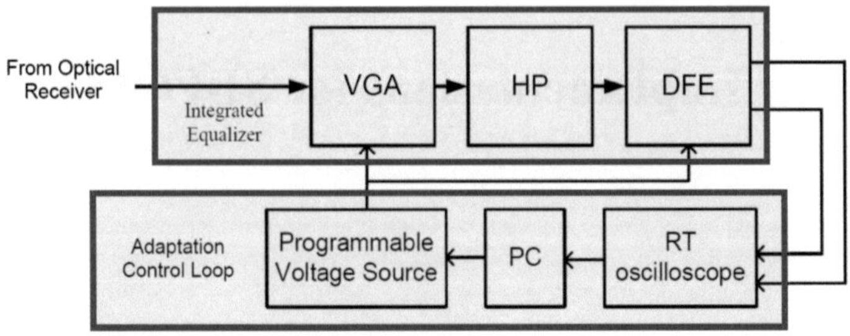

Fig. 8.1 Integrated equalizer and the external adaptation loop [11]

The performance can be enhanced more if an integrated equalizer designed for POF is used as in [11]. An integrated adaptive post-equalizer for SI-POF is needed. The first integrated POF equalizer was introduced in [11].

A discrete optical receiver (Hamamatsu S5052 800 μm PIN-PD with 28 V reverse bias and a Maxim 3266 TIA) was used. The output of the optical receiver was fed to an integrated equalizer chip. A data rate of 1.25 Gbit/s was achieved over 50 m SI-POF using a 650 nm RC-LED. The BER was 10^{-6}. Figure 8.1 shows the integrated equalizer and the external adaptation loop. The VGA acts as the only forward tap of the DFE portion; this tap is necessary to maintain a constant amplitude level for the equalized signal. The VGA is controlled via an analogue voltage. The high pass filter (HP) is adjusted via a 5-bit digital control word and can deliver a boost of 20 dB electrical amplitude at higher frequencies. The transfer function is controlled by the size of a source degeneration capacitor of a differential amplifier. The DFE features five feedback taps with continuously settable coefficients.

It samples the symbol and the sign of the error at the internal summing node and outputs them. The implementation has been done in the 180 nm CMOS technology of UMC.

The design optical receiver in [106] was tested for 1.25 Gb/s with an NRZ PRBS $(2^{31}-1)$ pattern over 50 m Mitsubishi GH SI-POF. The optical receiver composed of a discrete S5972 silicon PD from Hamamatsu (0.8 mm diameter, 0.44 A/W responsivity at 650 nm), TIA, equalizer and 50 Ω driver, see Fig. 8.2. A sensitivity of −8.2 dBm at BER = 10^{-12} was achieved in the 50 m POF at 1.25 Gb/s. The circuit power consumption was 96.8 mW for 50-m POF where the TIA consumes 18.6 mW, the line equalizer 3.8 mW, and the output driver 74.3 mW.

All of the above optical receivers used external components; the integrated equalizer reported in [11] still needed a separate optical receiver. The integrated optical receiver in [106] needs external photodiodes.

8.2 Fully Integrated Optical Receiver Containing an Equalizer

A PIN photodiode with 400 μm diameter is integrated within the presented optical receiver chip. The anode of the photodiode is formed by the p^+ substrate and is connected to ground, see Fig. 5.4. Two PIN-PDs are integrated with the optical receiver.

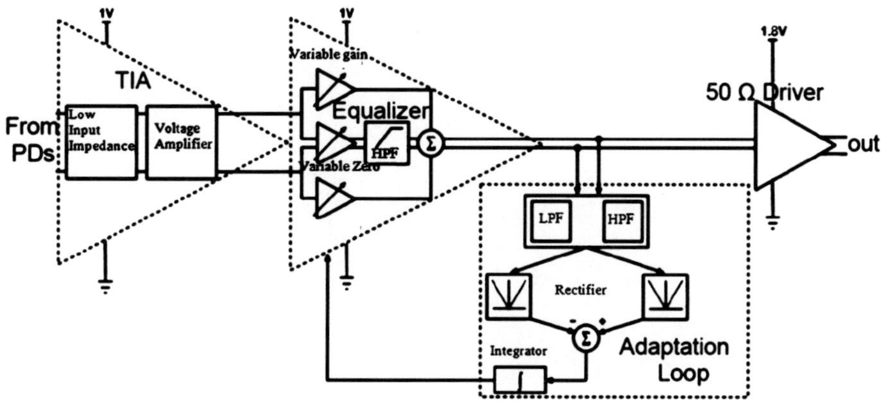

Fig. 8.2 Block diagram of the optical receiver with integrated equalizer [106]

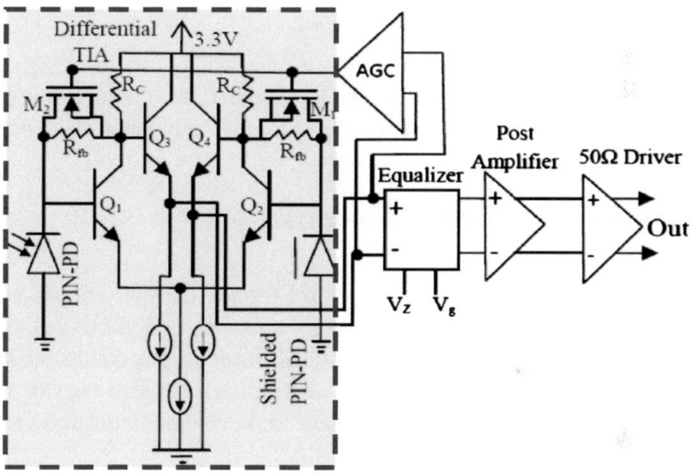

Fig. 8.3 Circuitry of the single-chip optical receiver containing an equalizer

One is active to collect the optical power and the other one is shielded to balance the differential transimpedance amplifier input. The differential topology is preferred for the TIA because of its high immunity against power supply and common-mode noise, see Fig. 8.3. An AGC sets the gain of the TIA to avoid overloading of the TIA and to increase the maximum received optical power to 1 mW (0 dBm).

The AGC makes the TIA operate in the linear region of operation to prevent the limiting effect of the TIA at high input optical power. This linear operation of the AGC-TIA is required for the POF equalizer in the next stage to work properly. The TIA has a −3 dB bandwidth of 622 MHz and an input referred noise current of about 170 nA. The outputs of the differential AGC-TIA are fed to the integrated POF equalizer. The output signal of the POF equalizer is amplified by the post amplifier

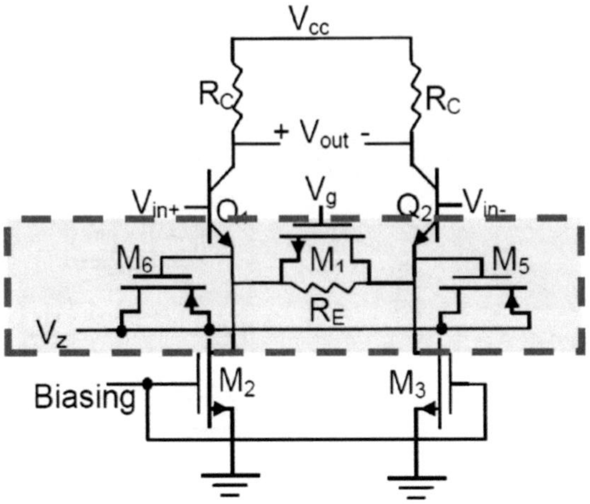

Fig. 8.4 Circuitry of the implemented equalizer cell

(Fig. 8.3), which is a limiting amplifier. The last stage in the optical receiver is a 50 Ω output driver which provides impedance matching to the measurement equipment.

8.2.1 The Integrated SI-POF Equalizer

The used equalizer in Fig. 8.4 has two control inputs, the first control signal (V_Z) controls the high pass filter corner frequency and the second control signal (V_g) controls the low-frequency gain. The high cut-off frequency is controlled by varying the MOS capacitors M_5 and M_6 via V_Z. The additional passive resistor R_E is used to give a proper minimum low-frequency gain to prevent an undesired too low low-frequency gain when M_1 is off [10, 64].

A similar equalizer in pure CMOS was presented by [41] to equalize for the copper cable limited bandwidth for gigabit transmission using several equalizer stages. Here we introduce a BiCMOS version with the bipolar transistors Q_1 and Q_2 resulting in a larger gain.

The values of the resistors R_C and R_E are selected to achieve a high gain. V_g is held constant to obtain the required low-frequency gain. So the effect of V_g on the low-frequency gain will be neglected and the only control is just done via V_Z. This gives a simple and effective control for the equalizer which is sufficient to equalize different POF lengths of up to 50 m.

8.2.2 Experimental Results for Two-Stage Equalizer

The presented equalizer has two cascaded cells of the equalizer cell presented in Fig. 8.4. The two-stage cascaded equalizer is integrated with the optical receiver

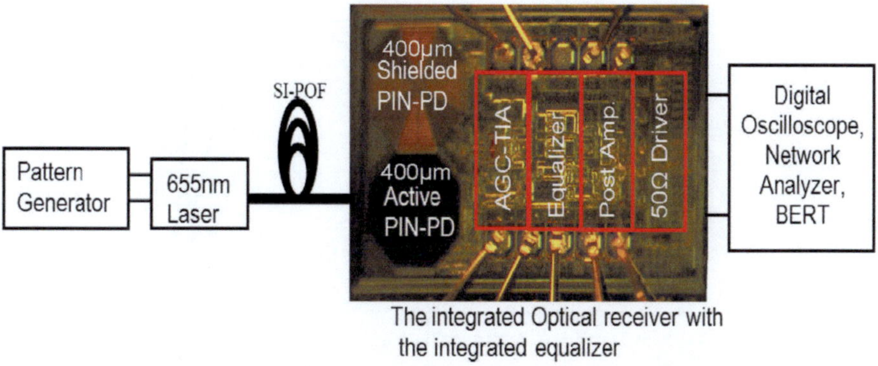

Fig. 8.5 Block diagram of the measurement setup

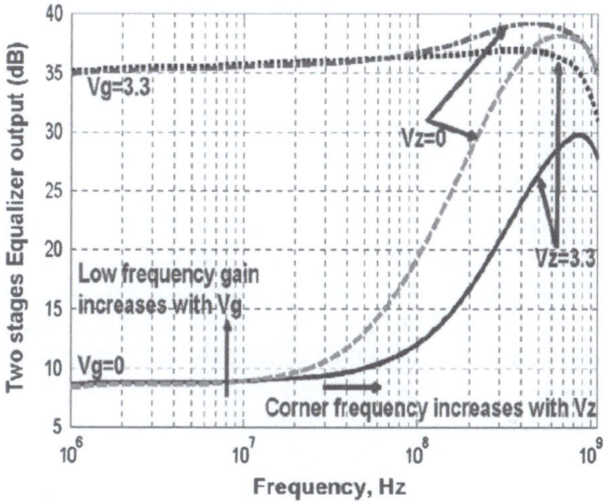

Fig. 8.6 Measured frequency response of the two-stage equalizer at different control signal values

presented in Fig. 8.3. The optical receiver uses a single 3.3 V supply and consumes 112 mW. The test chip is fabricated in 0.6 μm BiCMOS technology and occupies an area of 1.33 mm^2 [10]. To see the effect of the equalizer control signals on its frequency response; the network analyzer (HP 8753E) was used to modulate a 655 nm LD. Figure 8.5 shows the block diagram of the measurement setup. A glass optical fiber (GOF) was coupled to the 655 nm LD. The optical power after the GOF was received by the integrated optical receiver. The measured frequency response of the two-stage equalizer with different control signal values is shown in Fig. 8.6. Figure 8.6 illustrates the effect of the control signals V_z and V_g on the equalizer's low-frequency gain and on the corner frequency. By increasing V_g the low-frequency gain increases and by increasing V_z the corner frequency increases.

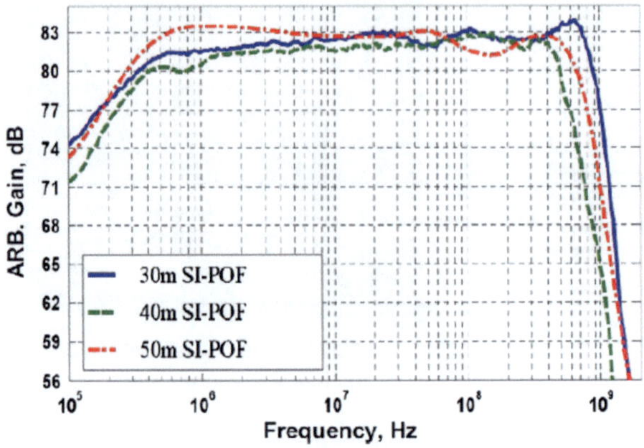

Fig. 8.7 Measured equalized frequency response at the output of the optical receiver at different POF lengths

Then the frequency response for the equalized SI-POF was measured. The network analyzer (HP 8753E) was used to modulate a 655 nm LD. The PMMA SI-POF (Luceat POF1.100.B22M) with a core diameter of 1 mm and NA $= 0.5$ was coupled to the 655 nm LD. The optical power after the POF was received by the presented integrated optical receiver using butt coupling. The coupling loss from the 1 mm POF to the 400 μm photodiode was -7 dB. The measured frequency response for different SI-POF lengths (30, 40 and 50 m) using the presented optical receiver is shown in Fig. 8.7. It is clear from Fig. 8.7 that the presented integrated equalizer succeeded to equalize POF lengths up to 50 m to have enough bandwidth (600 MHz) for a data rate of 1 Gbit/s. For sensitivity measurements the 655 nm LD was modulated by a 1 Gbit/s binary signal at PRBS $= 2^7 - 1$ and 4.5 dBm optical power was coupled into the SI-POF. The measured BER and the received average optical power at different POF lengths are illustrated in Fig. 8.8. Up to 40 m POF the received signal is error free (BER $= 10^{-9}$) with -14 dBm average received optical power (collected by the optical receiver); so there is a 6.7 dB optical power margin. The measured BER for 50 m POF was 10^{-6} with -14 dBm average received optical power (collected by the optical receiver) resulting in 5.5 dBm power margin. A summary for the measured sensitivities, received optical power and power margin for different POF lengths is included in Table 8.1.

The integrated equalizer succeeds to compensate for the SI-POF's limited bandwidth for different POF lengths 20, 30, 40, and 50 m to achieve enough bandwidth for data rates up to 1 Gbit/s. It is clear from the equalized frequency response illustrated in Fig. 8.7 that the frequency responses for the different equalized POF lengths are not the same. For the 30 and 40 m POF the frequency response is flat, whereas the frequency response for the equalized 50 m POF has a 2 dB undershoot between 100 and 200 MHz. The integrated two-stage equalizer cannot reach an optimum compensation for the 50 m POF. This is why the transmission of 1 Gbit/s over 50 m

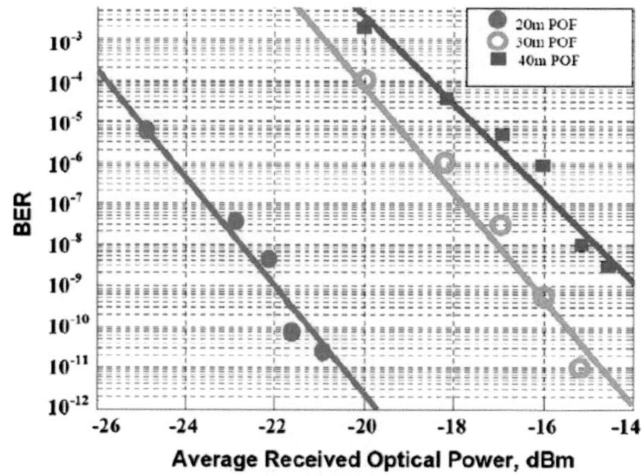

Fig. 8.8 BER for 1 Gbit/s with PRBS $= 2^7 - 1$ binary data transmission over 20, 30 and 40 m PMMA SI-POF

Table 8.1 A summary for the measured received optical power, sensitivities, and power margin for different POF lengths

POF length	20 (m)	30 (m)	40 (m)	50 (m)
Received average power (dBm)	−4.9	−6.1	−7.3	−8.5
Sensitivity (BER $= 10^{-9}$) (dBm)	−22	−16.5	−14	−14 (BER $= 10^{-6}$)
Power margin (dB)	17.1	10.4	6.7	5.5
	145			

POF cannot reach a BER less than 10^{-6}, whereas the 20, 30, and 40 m POF lengths allow reaching a BER of 10^{-9}. There is a 5.5 dB optical power margin for 1 Gbit/s over 50 m POF (Table 8.1) and it can be increased to 8.5 dB by using lens coupling to the integrated 400 μm PIN photodiode. This large optical power margin enables us to migrate to a 1 mW VCSEL (for eye safety) with 5 dB optical power margin.

8.2.3 Experimental Results for Single-Stage Equalizer

The presented equalizer has one cell from the equalizer cell presented in Fig. 8.4. The single-stage equalizer is integrated with the same optical receiver presented in Fig. 8.3. The optical receiver containing a single-stage equalizer uses a single 3.3 V supply and consumes 100 mW. The test chip is fabricated in 0.6 μm BiCMOS technology and occupies an area of 1.31 mm^2 [64]. Die photo of the complete optical receiver is included in Fig. 8.5.

For POF applications, a single-stage equalizer will be enough for POF lengths up to 50 m. This is because the POF lengths up to 50 m show a frequency response similar

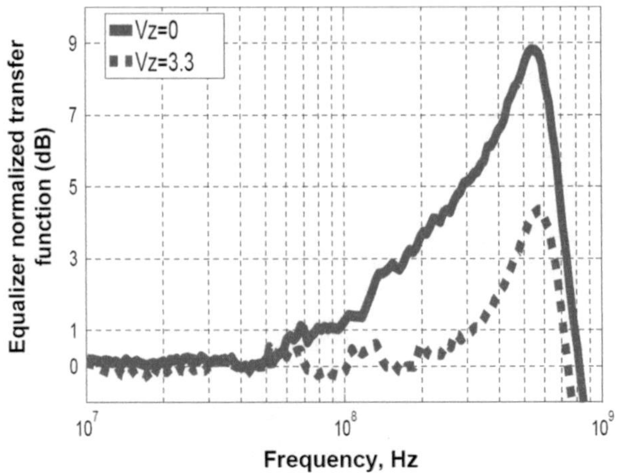

Fig. 8.9 Measured transfer function of the optical receiver containing a single-stage equalizer for different control signal values

to a first-order low-pass filter (see Sect. 4.3.1, Table 4.2). The measured transfer function of the optical receiver containing a single-stage equalizer (shown in Fig. 8.4) is illustrated in Fig. 8.9.

It is clear from Fig. 8.9 that by increasing V_Z the corner frequency increases. This high pass filter effect of the equalizer can compensate for the low pass filter behavior of the SI-POF. In the presented design we aimed to reduce the number of control signals from two to one (only V_Z).

For the frequency response measurements a network analyzer (HP 8753E) was used to modulate a 655 nm LD. The PMMA SI-POF with a core diameter of 1 mm and NA = 0.5 was butt coupled to the 655 nm LD. The optical power after the POF was received by the integrated optical receiver also using butt coupling. The block diagram for the measurement setup is shown in Fig. 8.5. The output electrical power of the receiver measured as function of frequency for different equalized SI-POF lengths (20 and 50 m) using the presented optical receiver is shown in Fig. 8.10. The received electrical power is proportional to the square of the optical power, due to the square-law effect of the PD. So, the −3 dB optical bandwidth will be the −6 dB electrical bandwidth. Together with the transmitter and the optical receiver, the 20 m POF has an equalized bandwidth of 750 MHz and a 50m POF results in an equalized bandwidth of 700 MHz. This bandwidth is sufficient for giga bit transmission over SI-POF.

For BER measurement the 655 nm LD was modulated by a 1.8 Gbit/s or a 1.25 Gbit/s signal at PRBS = $2^{15} - 1$ and 5 dBm optical power was butt coupled to a 20 and a 50 m SI-POF, respectively.

The received data stream from the integrated optical receiver was compared with the original transmitted signal by bit error counting using a bit error rate tester. The measured BER in dependence on the received optical power at different equalized

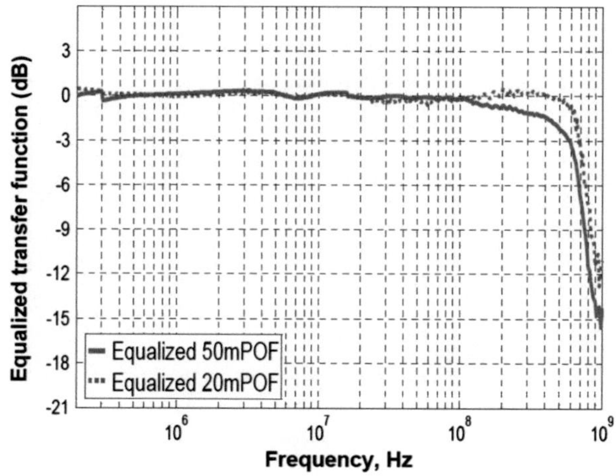

Fig. 8.10 Transfer function of optical receiver with single-stage equalizer for 20 and 50 m SI-POF

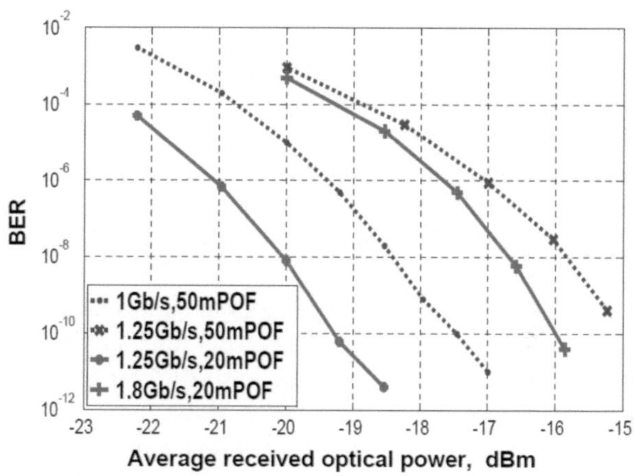

Fig. 8.11 Measured BER for the equalized POF

POF lengths are illustrated in Fig. 8.11. 1.25 and 1.8 Gbit/s can be transmitted over 20 m SI-POF with sensitivity of -19.6 and -16.5 dBm, respectively, with a BER of 10^{-9} and a PRBS with a length of $2^{15} - 1$.

Data rates of 1 and 1.25 Gbit/s can be transmitted over 50 m SI-POF with sensitivity of -18 and -15.5 dBm, respectively, at a BER of 10^{-9} and a PRBS with the length of $2^{15} - 1$. The average optical power received by the integrated photodiode is measured by an on-chip circuitry (actually the photocurrent is measured and the optical power is calculated with the known responsivity (0.52 A/W at 660 nm) of the integrated photodiode).

The transmitted average optical power from the 655 nm LD is 5 dBm. The maximum average optical power received by the optical receiver after 50 m POF is -8 dBm.

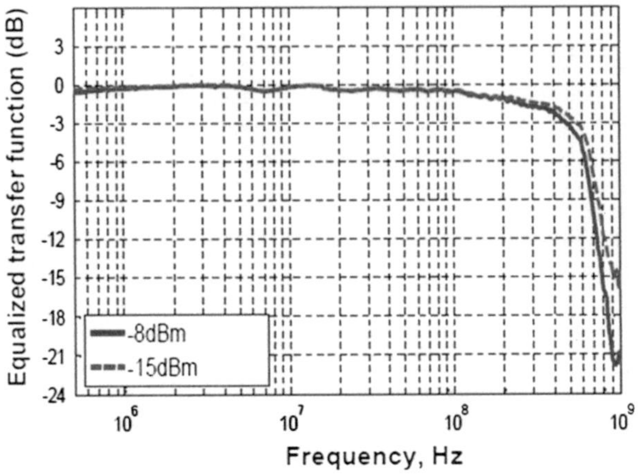

Fig. 8.12 Measured frequency response of the optical receiver containing the single-stage equalizer after 50 m POF for different received average optical power

The 13 dB loss from the LD to the integrated optical receiver comes from the POF attenuation (7 dB after 50 m) and the coupling loss of the 1 mm core diameter POF to the 0.4 mm diameter PIN-PD (6 dB).

By using the presented single-chip optical receiver with the integrated equalizer a sensitivity of −15.5 dBm for 1.25 Gbit/s transmitted over 50 m POF is achieved. There will be a 7.5 dB optical power margin (15.5 − 8 dBm = 7.5 dBm) for 1.25 Gbit/s transmission over 50 m SI-POF.

The transfer function of the equalized 50 m SI-POF is shown in Fig. 8.12 for different optical power values (−15 and −8 dBm). The transfer functions in Figs. 8.10 and 8.12 indicate that the integrated equalizer works successfully for gigabit transmission over 50 m SI-POF and at different input optical power levels.

The measured eye diagrams after 20 m SI-POF with a data rate of 1.8 Gbit/s, PRBS $2^{15}-1$ and a received average optical power of −16 and −8 dBm are illustrated in Fig. 8.13a, b, respectively. The measured eye diagrams after 50 m SI-POF with a data rate of 1.25 Gbit/s, PRBS $2^{15}-1$ and a received average optical power of −16 and −8 dBm are illustrated in Fig. 8.13c, d, respectively. It is clear from Fig. 8.13 that 1.25 Gbit/s transmission over 50 m POF and 1.8 Gbit/s transmission over 20 m POF is possible with high sensitivity and at different input optical power levels.

A data rate of 1.25 Gbit/s was transmitted over 50 m SI-POF with three circular bendings in the POF near to the laser source with a radius of 12 mm. The optical signal from the POF was received and equalized by the integrated optical receiver. The output eye diagram of the equalized signal is shown in Fig. 8.14. The bending increases the modal dispersion effect because more modes are exited and therefore the system bandwidth is reduced and the jitter is increased, see Fig. 8.14. It is clear from the eye diagram in Fig. 8.14 that the effect of POF bending still can be equalized to make 1.25 Gbit/s over 50 m POF possible.

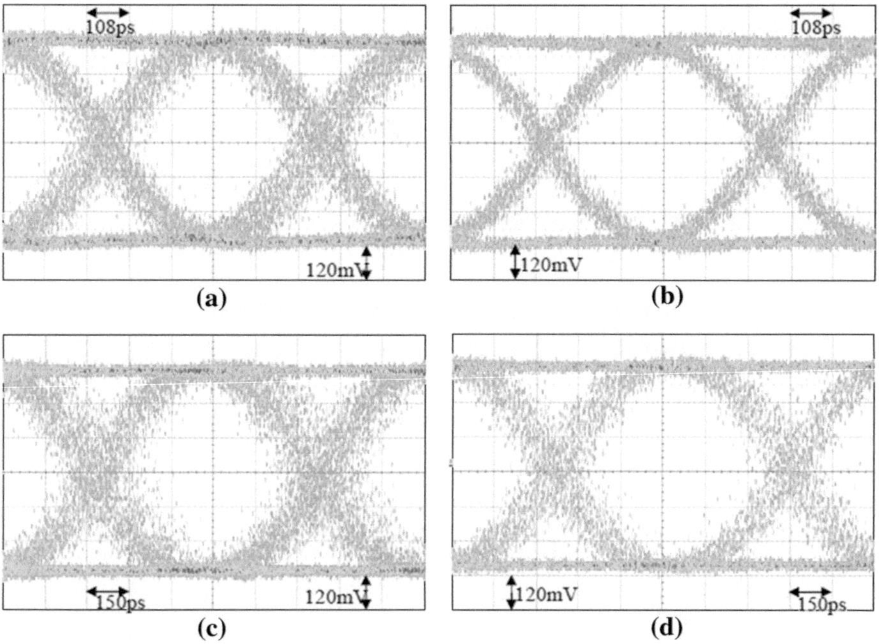

Fig. 8.13 Eye diagrams measured at the output of the integrated optical receiver with the integrated equalizer for a binary signal with PRBS $= 2^{15} - 1$ at: **a** 1.8 Gbit/s over 20 m SI-POF with an average received optical power of −16 dBm, **b** 1.8 Gbit/s over 20 m SI-POF with an average received optical power of −8 dBm, **c** 1.25 Gbit/s over 50 m SI-POF with an average received optical power of −16 dBm, **d** 1.25 Gbit/s over 50 m SI-POF with an average received optical power of −8 dBm

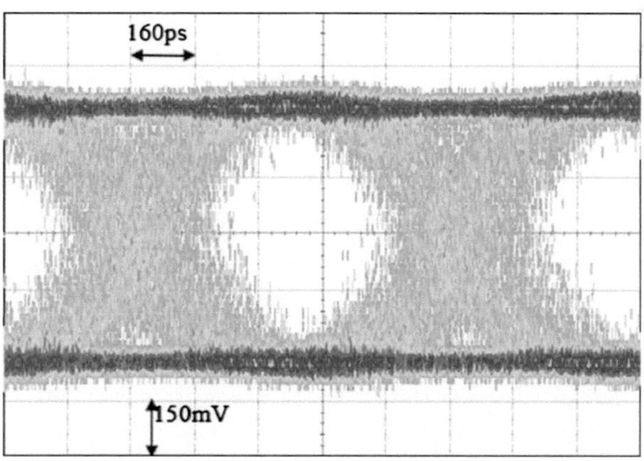

Fig. 8.14 Measured eye diagrams for a data rate of 1.25 Gbit/s with PRBS $= 2^{15} - 1$ over 50 m SI-POF with three circular bendings with 12 mm bending radius

Table 8.2 Summary of integrated optical receivers with integrated equalizer for application with SI-POF

	POF length (m)	DR (Gb/s)	Sensitivity (dBm)	Power (mW)	Technology	FOM
[64]	50	1.25	−15.5	100	0.6 μm BiCMOS	9.7
[106]	50	1.25	−8.2	96.78	180 nm CMOS	5.3
[107]	50	1	−15.5*	165	0.35 μm CMOS	4.7

* Assumed, since the sensitivity was not reported

8.3 Summary of Integrated Optical Receivers with Integrated SI-POF Equalizer

To compare with other integrated optical receiver solutions, a possible figure-of-merit (FOM) introduced in [106] is the following:

$$\mathrm{FOM} = \frac{\mathit{Data\ Rate}(\mathrm{Gb/s}) \cdot \mathit{POF\ Length}(\mathrm{m}) \cdot \mathit{Sensitivity}(\mathrm{dBm})}{\mathit{Power\ Consumption}(\mathrm{mW})} \tag{8.1}$$

By using this FOM, the 0.6 μm BiCMOS fully integrated optical receiver introduced in [64] (shown in Fig. 8.3) provides the best FOM of 9.7. The integrated optical receiver fabricated in 180 nm CMOS technology with discrete photodiodes presented in [106] has a FOM of 5.3. An analog equalizer proposed in [107] targets 1 Gb/s through 50 m SI-POF with power consumption of 165 mW and assuming the same sensitivity of −15.5 dBm as in [64] (not available). This is providing a FOM of 4.7. The summary of integrated optical receivers with integrated SI-POF equalizer is shown in Table 8.2.

Chapter 9
Conclusions

The standard PMMA SI-POF is very attractive for use in short-range communication and in-building networks. Due to modal dispersion, the SI-POF's bandwidth is small and limits the maximum data rate which can be transmitted. To overcome the problem of the PMMA POF's high transmission loss and low bandwidth we introduced a sensitive receiver with a high linearity for multilevel signaling.

The application of M-PAM modulation and equalization techniques in a PMMA SI-POF channel for short-range optical communications was proposed and investigated in this book. The reason to investigate the use of M-PAM is because of its higher spectral efficiency than that of a binary signal. Naturally, M-PAM has disadvantages of an extra power penalty and of a higher implementation complexity compared to binary signals, which were also investigated in this book.

The principle of M-PAM was explained and it was shown that the use of 4-PAM has a benefit over binary signal when the channel frequency has a roll-off larger than 5.1 dB/octave. Furthermore, the standard SI-POF's frequency response was measured for different POF lengths and it was proven that in the case for SI-POF with lengths longer than 40 m 4-PAM has a better performance than binary signaling.

The main target of the work described in this book was the design of fully integrated optical receiver to achieve a high transmission performance over SI-POF. Three types of integrated optical receivers having a high linearity with multilevel signaling were designed, fabricated and characterized. The first integrated optical receiver has a bandwidth of 622 MHz and a sensitivity of −24 dBm at a DR of 1.25 Gb/s. The second one is a 200 MHz bandwidth optical receiver with −29.5 dBm sensitivity at 400 Mb/s DR. The last integrated optical receiver has a bandwidth of 112 MHz and a sensitivity of −31 dBm at a DR of 250 Mb/s.

Experiments regarding the application of 4-PAM and 8-PAM over SI-POF channel were presented using the fabricated linear integrated optical receivers. The most interesting results are the ability to transmit data at a rate of 1 Gbit/s over 20 m SI-POF, 400 Mb/s over 50 m SI-POF and 170 Mbit/s over 115 m of SI-POF. All of the reached results were free of error ($<10^{-8}$), were achieved in real time and have a data rate-length product of 20 Gb/s·m. The main advantages in our system are the use

M. Atef and H. Zimmermann, *Optical Communication over Plastic Optical Fibers*, Springer Series in Optical Sciences 172, DOI: 10.1007/978-3-642-30388-3_9,

of M-PAM modulation with an integrated optical receiver with a high performance with M-PAM in linearity and sensitivity.

Two different equalizers were implemented on chip together with pin photodiode, TIA, post amplifier and output driver to investigate also binary data transmission over POF with long lengths. The first one consisted of two identical cascaded equalizer stages. The second one contained only one equalizer stage. Both integrated optical receivers were fully integrated also with the equalizers. A single-chip optical receiver with an integrated equalizer was used to achieve a high step-index plastic optical fiber (SI-POF). The integrated equalizer can compensate for different POF lengths up to 50 m. The integrated optical receiver with on-chip equalizers were fabricated in a low-cost silicon 0.6 μm BiCMOS technology and had a power consumption of about 100 mW. Real-time transmission at data rates of 1.8 Gbit/s over 20 m SI-POF and 1.25 Gbit/s in binary mode over 50 m SI-POF with high sensitivities and low BER was achieved.

In summary, although for red light (650 nm) the POF's attenuation is quite large, it was shown that integrated optical receivers with M-PAM or preferably with binary modulation using an integrated equalizer can extend the POF's length to practically interesting and usable lengths for data rates in excess of 1.25 Gbit/s.

References

1. ETSI TS 105 175–1 V1.1.1(2010–01), Access, Terminals, Transmission and Multiplexing (ATTM); Plastic Optical Fibre System Specifications for 100 Mbit/s and 1 Gbit/s, http://www.etsi.org/WebSite/homepage.aspx. Accessed January 2010
2. P. Polishuk, Plastic optical fibers branch out. IEEE Commun. Mag. **44**(9), 140–148 (2006)
3. R. Gaudino, E. Capello, G. Perrone, G. Perrone, M. Chiaberge, P. Francia, G. Botto, Advanced modulation format for high speed transmission over standard SI-POF using DSP/FPGA platforms, in *POF Conference 2004*, Nuremberg, Sept 2004, pp. 98–105
4. F. Breyer, S. Lee, S. Randel, N. Hanik, PAM-4 signalling for gigabit transmission over standard step-index plastic optical fibre using light emitting diodes, in *34th European Conference and Exhibition on Optical Communication (ECOC 2008)*, vol. 3, Brussels, Belgium, Sept 2008, pp. 81–82
5. S.C.J. Lee, F. Breyer, D. Cardenas, S. Randel, A.M.J. Koonen, Real-time gigabit DMT transmission over plastic optical fibre. Electron. Lett. **45**(25), 1342–1343 (2009)
6. S.C.J. Lee, F. Breyer, S. Randel, R. Gaudino, G. Bosco, A. Bluschke, M. Matthews, P. Rietzsch, H.P.A. van den Boom, A.M.J. Koonen, Discrete multitone modulation for maximizing transmission rate in step-index plastic optical fibers. J. Lightwave Technol. **27**(11), 1503–1513 (2009)
7. O. Ziemann, L. Giehmann, P.E. Zamzow, H. Steinberg, D. Tu, Potential of PMMA Based SI-POF for Gbps transmission in automotive applications, in *The 9th International POF Conference*, Cambridge, Oct 2000, pp. 44–48
8. F. Breyer, N. Hanik, S. Randel, B. Spinnler, Investigations on electronic equalization for step-index polymer optical fiber systems, in *Proceedings Symposium IEEE/LEOS Benelux Chapter*, Eindhoven, Nov/Dec 2006, pp. 149–152
9. F. Breyer, S. Lee, S. Randel, N. Hanik, 500-Mbit/s transmission over 50 m standard 1-mm step-index polymer optical fiber using PAM4-modulation and simple equalization schemes, in *Proceedings of Photon One Summer School*, Brest, France, July, 2007
10. M. Atef, R. Swoboda, H. Zimmermann, 1 Gbit/s transmission over step-index plastic optical fiber using an optical receiver with an integrated equalizer. Opt. Commun. **284**(21), 5153–5156 (2011)
11. C. Zerna, J. Sundermeyer, A. Fiederer, N. Verwaal, B. Offenbeck, N. Weber, Integrated PAM2 decision feedback equalizer for gigabit ethernet over standard SI-POF using red LED, in *36th European Conference on Optical Communication (ECOC)*, 2010, p. We.6.B.4
12. K. Farzan, D.A. Johns, A robust 4-PAM signaling scheme for inter-chip links using coding in space. IEEE Trans. VLSI Syst. **16**(11), 1535–1544 (2008)

M. Atef and H. Zimmermann, *Optical Communication over Plastic Optical Fibers*, Springer Series in Optical Sciences 172, DOI: 10.1007/978-3-642-30388-3,

13. G. Schroeder, Addressing emerging test challenges for multilevel signaling devices, in *International Electronics Manufacturing Technology Symposium* (*IEMT*), San Jose, CA, July 2002, pp. 376–381
14. S. Walklin, J. Conradi, Multilevel signaling for increasing the reach of 10 Gb/s lightwave systems. J. Lightwave Technol. **17**, 2235–2248 (1999)
15. J.E. Cunningham, D. Beckmanhy, X. Zheng, D. Huang, T. Sze, A.V. Krishnamoorthy, PAM-4 signaling over VCSELs with 0.1 3Âμm CMOS chip technology. Opt. Express **14**(25), 12028–12038 (2006)
16. J.K. Pollard, Multilevel data communication over optical fibre. IEE Proc. I Commun. Speech Vis. **138**(3), 162–168 (1991)
17. K.P. Ho, *Phase-Modulated Optical Communication Systems* (Springer, New York, 2005)
18. K. Yamaguchi, K. Sunaga, S. Kaeriyama, T. Nedachi, M. Takamiya, K. Nose, Y. Nakagawa, M. Sugawara, M. Fukaishi, 12 Gb/s duobinary signaling with x2 oversampled edge equalization, in *International Solid-State Circuits Conference* (*ISSCC*), vol. 1 (2005), pp. 70–71
19. J.M. Kahn, K.P. Ho, T.M. Hollis, Review of advanced modulation and signal processing techniques for 40 Gb/s optical transmission systems. Online lecture notes, http://www.doe.carleton.ca/lzhang/OpticsFinal.ps, June 2005, pp. 1–12
20. S. Benedetto, E. Biglieri, *Principles of Digital Transmission with Wireless Applications* (Kluwer/Plenum, New York, 1999)
21. J.G. Proakis, M. Salehi, *Fundamentals of Communication Systems* (Pearson Prentice Hall, Upper Saddle River, 2005)
22. J.G. Proakis, *Digital Communications*, 4th edn. (McGraw-Hill, New York, 2001)
23. S. Randel, F. Breyer, S. Lee, High-speed transmission over multimode optical fibers, in *Optical Fiber Communication and the National Fiber Optic Engineers Conference* (*OFC/NFOEC 2008*), San Diego, CA, USA, Feb 2008, p. OWR2
24. S. Jung, J. Gao, CMOS Multi-level signal transmitter for optical communication, in *The 47th IEEE International Midwest Symposium on Circuits and Systems*, vol. 2, July 2004, pp. II-185–II-188
25. http://www.acquitek.com
26. http://www.sympuls-aachen.de
27. M. Meghelli, A. Konczykowska, J. Godin, P. Desrousseaux, P. Andre, Multilevel decoder-decision circuit for high bitrate ETDM transmission, in *European Solid-State Circuits Conference* (*ESSCIRC 1997*), Sept 1997, pp. 368–371
28. A.X. Widmer, P.A. Franaszek, A DC-balanced, partitioned-block, 8B/10B transmission code. IBM J. Res. Dev. **27**(5), 440–451 (1983)
29. S. Walklin, J. Conradi, Multilevel signaling for extending the dispersion-limited transmission distance in high-speed, fiber optic communication systems, in *Conference on Electrical and Computer Engineering*, vol. 1 (1996), pp. 233–236
30. S. Hranilovic, D.A. Johns, A multilevel modulation scheme for high-speed wireless infrared communications, in *IEEE International Symposium on Circuits and Systems* (*ISCAS*), vol.6 (1999), pp. 338–341
31. R. Gaudino, D. Cardenas, P. Spalla, A. Nespola, S. Abrate, Multilevel 50 Mb/s transmission over a 200 m PMMA SI-POF LAN testbed, in *14th International Conference on Polymer Optical Fiber (POF 2005)*, Hong Kong, Sept 2005, pp. 207–210
32. A. Nespola, S. Camatel, S. Abrate, D. Cardenas, R. Gaudino, Equalization techniques for 100 Mb/s data rates on SI-POF for optical short reach applications. in *Optical Fiber Communication and the National Fiber Optic Engineers Conference* (*OFC/NFOEC 2007*), March 2007, p. OMR1
33. F. Breyer, S. Lee, D. Cardenas, S. Randel, N. Hanik, Real-time gigabit ethernet transmission over up to 25 m step-index polymer optical fibre using LEDs and FPGA-based signal processing, in *35th European Conference on Optical Communication* (*ECOC 2009*), Vienna, Austria, Sept 2009, p. Paper 3.5.5

34. M.S. Filip Tavernier, *High-Speed Optical Receivers with Integrated Photodiode in Nanoscale CMOS* (Springer, New York, 2011)
35. M. Kiziroglou, A. Mukherjee, S. Vatti, A. Holmes, C. Papavassiliou, E. Yeatman, Self-assembly of three-dimensional Au inductors on silicon. IET Microw. Antennas Propaga. **4**(11), 1698–1703 (2010)
36. J.M. Cioffi, T. Kailath, Fast, recursive-least-squares transversal filters for adaptive filtering. IEEE Trans. Acoust. Speech Signal Process. **32**(2), 304–337 (1984)
37. H. Kim, F. Bien, Y. Hur, S. Chandramouli, J. Cha, E.Gebara, J. Laskar, A 0.25-μm BiCMOS feed forward equalizer using active delay line for backplane communication, in *ISCAS*, 2007, pp. 193–196
38. M. Maeng, F. Bien, Y. Hur, H. Kim, 0.18 um CMOS equalization techniques for 10-Gb/s fiber optical communication. IEEE Trans. Microw. Theory Tech. **53**(11), 3509–3519 (2005)
39. D. Hernandez-Garduno, J. Silva-Martinez, A CMOS 1 Gb/s 5-tap fractionally-spaced equalizer. IEEE J. Solid-State Circuits **43**(11), 2482–2491 (November 2008)
40. H. Wang, J. Lee, A 21-gb/s 87-mw transceiver with FFE/DFE/analog equalizer in 65-nm CMOS technology. IEEE J. Solid-State Circuits **45**(4), 909–920 (2010)
41. J.-S. Choi, M.-S. Hwang, D.-K. Jeong, A 0.18 μm CMOS 3.5-Gb/s continuous-time adaptive cable equalizer using enhanced low-frequency gain control method. IEEE J. Solid-State Circuits **39**(3), 419–425 (2004)
42. M.H. Hayes, *Statistical Digital Signal Processing and Modeling* (Wiley, New York, 1996)
43. J.K. Omura, On the Viterbi decoding algorithm. IEEE Trans. Inf. Theory **15**(1), 177–179 (1969)
44. G.D. Forney, Convolutional codes II. Maximum-likehood decoding. Inf. Control **25**(3), 222–266 (1974)
45. KDPOF, Demonstration of 1 Gbps over 50 m of Low Cost SI-POF with KDPOF Technology, http://www.kdpof.com, June 2010, pp. 1–5
46. G. Jiang, R.F. Shi, A.F. Garito, Mode coupling and equilibrium mode distribution conditions in plastic optical fibers. IEEE Photon. Technol. Lett. **9**(8), 1128–1130 (1997)
47. J. Zubia, J. Arrue, Plastic optical fibers: an introduction to their technological processes and applications. J. Opt. Fiber Technol. **7**, 101–140 (2001)
48. J. Mateo, M.A. Losada, J. Zubia, Frequency response in step index plastic optical fibers obtained from the generalized power flow equation. Opt. Express **17**(4), 2850–2860 (2009)
49. O. Ziemann, J. Krauser, P.E. Zamzow, W. Daum, *POF Handbook—Optical Short Range Transmission Systems*, 2nd edn. (Springer, Berlin, 2008)
50. O. Ziemann, H. Poisel, M. Luber, M. Bloos, A. Bachmann, Optical data communication over short distances, state of the art of fiber types, transmitters and receivers. Laser + Photonik J. 36–41 (2005)
51. T. Numai, *Fundamentals of Semiconductor Lasers* (Springer, Berlin, 2004)
52. J. L. Miller, E. Friedman, *Optical Communications Rules of Thumb* (McGraw-Hill, Burr Ridge, 2002)
53. B. Chomycz, *Planning Fiber Optics Networks* (McGraw-Hill, New York, 2008)
54. S.C.J. Lee, H.P.A. van den Boom, R.L. Duijn, P. van Bennekom, J. Zeng, S. Randel, B. Spinnler, A.M.J. Koonen, 1.25 Gb/s transmission over 25 m of 1 mm standard step-index Pmma polymer optical fiber for gigabit ethernet, in *11th European Conference on Networks and Optical Communications* (*NOC 2006*), Berlin, July 2006, pp. 110–117
55. E. Grivas, N. Raptis, D. Syvridis, An optical mode filtering technique for the improvement of the large core SI-POF link performance. J. Lightwave Technol. **28**, 1796–1801 (2010)
56. S.C.J. Lee, F. Breyer, D. Cardenas, S. Randel, T. Koonen, Real-time implementation of a 1.25-Gbit/s DMT transmitter for robust and low-cost LED-based plastic optical fiber applications, in *35th European Conference and Exhibition on Optical Communication* (*ECOC 2009*), Vienna, Austria, Sept 2009, p. Paper 3.5.4

57. S.C.J. Lee, H.P.A. van den Boom, S. Randel, D. Cardenas, R.L. Duijn, S. Randel, B. Spinnler, P. van Bennekom, F.M. Huijskens, J. Zeng, A.M.J. Koonen, G.D. Khoe, 2 × 500 Mb/s transmission over 25 m of standard 1 mm step-index PMMA POF with angular mode group diversity multiplexing, in *Proceedings of the 15th International Conference on Plastic Optical Fibers (POF 2006)*, Seoul, Korea, July 2006, pp. 348–353
58. S. Randel, S. Lee, B. Spinnler, F. Breyer, H. Rohde, J. Walewski, A.M.J. Koonen, A. Bluschke, 1 Gbit/s transmission with 6.3 bit/s/Hz spectral efficiency in a 100 m standard 1 mm step-index plastic optical fibre link using adaptive multiple sub-carrier modulation, in *32th European Conference and Exhibition on Optical Communication (ECOC 2006)*, Cannes, France, Post dead line Paper Th4.4.1, Sept 2006
59. B. Charbonnier, P. Urvoas, M. Ouzzif, J.L. Masson, J.D. Lambkin, M. O'Gorman, R. Gaudino, EU project POF-PLUS: gigabit transmission over 50 m of step-index plastic optical fibre for home networking, in *Optical Fiber Communication and the National Fiber Optic Engineers Conference (OFC/NFOEC 2009)*, March 2009, p. OWR4
60. M. Atef, R. Swoboda, H. Zimmermann, Gigabit transmission over PMMA step-index plastic optical fiber using an optical receiver for multilevel communication, in *36th European Conference and Exhibition on Optical Communication (ECOC)* (Torino, Italy, 2010) p. 6.02
61. M. Atef, W. Gaberl, R. Swoboda, H. Zimmermann, An integrated optical receiver for multilevel data communication over plastic optical fiber. Analog Integr. Circuits Signal Process. J. **67**(1), 3–9 (2011)
62. M. Atef, R. Swoboda, H. Zimmermann, An optical receiver for eight-level data communication over step index plastic optical fiber. J. Opt. Commun. **283**(11), 2350–2352 (2010)
63. A. Antonino, S. Straullu, S. Abrate, A. Nespola, P. Savio, D. Zeolla, J.C.R. Molina, R. Gaudino, S. Loquai, J. Vinogradov, Real-time gigabit ethernet bidirectional transmission over a single SI-POF up to 75 Meters, in *OSA/OFC/NFOEC*, 2011, p. OWA2
64. M. Atef, R. Swoboda, H. Zimmermann, 1.25 Gbit/s over 50 m step-index plastic optical fiber using a fully integrated optical receiver with an integrated equalizer. J. Lightwave Technol. **30**(1), 118–122 (2012)
65. A. Nespola, S. Straullu, P. Savio, D. Zeolla, J.C.R. Molina, S. Abrate, R. Gaudino, A new physical layer capable of record gigabit transmission over 1 mm step index polymer optical fiber. J. Lightwave Technol. **28**(20), 2944–2950 (2010)
66. D. Cardenas, A. Nespola, P. Spalla, S. Abrate, R. Gaudino, A media converter prototype for 10-Mb/s ethernet transmission over 425 m of large-core step-index polymer optical fiber. J. Lightwave Technol. **24**(12), 4946–4952 (2006)
67. D. Cardenas, A. Nespola, S. Camatel, S. Abrate, R. Gaudino, 100 Mb/s ethernet transmission over 275 m of large core step index polymer optical fiber: results from the Pof-ALL European project. J. Lightwave Technol. **27**(14), 2908–2915 (2009)
68. M. Atef, R. Swoboda, H. Zimmermann, 170 Mbit/s multilevel transmission over 115 m standard step-index plastic optical fiber using an integrated optical receiver. Opt. Commun. **284**(1), 191–194 (2011)
69. H. Zimmermann, *Integrated Silicon Optoelectronics* (Springer, Berlin, 2000)
70. H. Zimmermann, K. Schneider, *Highly Sensitive Optical Receivers* (Springer, New York, 2006)
71. M. Förtsch, H. Zimmermann, W. Einbrodt, K. Bach, H. Pless, Integrated PIN photodiodes in high-performance BiCMOS technology, in *Technical Digest of the International Electron Devices Meeting IEDM*, San Francisco, Dec 2002, pp. 801–804
72. M. Förtsch, H. Dietrich, H. Zimmermann, Complete low-cost 625 Mbit/s optical fiber receiver in 0.6 Âμm BiCMOS technology. Proc. SPIE Int. Soc. Opt. Eng. **5952**, 59520R1–59520R6 (2005)
73. M. Förtsch, H. Zimmermann, H. Pless, 220-MHz monolithically integrated optical sensor with large-area integrated PIN photodiode. IEEE Sens. J. **6**(2), 385–390 (2006)

74. P.J.-W. Lim, A.Y.C. Tzeng, H.L. Chuang, S.A.S. Onge, A 3.3-V monolithic photodetector/CMOS-preamplifier for 531 Mb/s optical data link applications, in *International Solid-State Circuits Conference* (*ISSCC*), 1993, pp. 96–97
75. M.K.M. Yamamoto, K. Nakao, Si-OEIC with a built-in PIN photodiode. IEEE Trans. Electron Devices **42**(1), 58–63 (1995)
76. C.T. Kirk, A theory of transistor cutoff frequency (Ft) falloff at high current densities. IRE Trans. Electron Dev. **9**, 164–174 (1962)
77. M. Ingels, G. Van Der Plas, J. Crols, M. Steyaert, A CMOS 18 THz Omega; 248 Mb/s transimpedance amplifier and 155 Mb/s LED-driver for low cost optical fiber links. IEEE J. Solid-State Circuits **29**(12), 1552–1559 (1994)
78. M. Atef, H. Zimmermann, 2.5 Gbit/s transimpedance amplifier using noise cancelling for optical receivers, in *ISCAS*, 2012
79. T. Vanisri, C. Toumazou, Integrated high frequency low-noise current-mode optical transimpedance preamplifiers: theory and practice. IEEE J. Solid-State Circuits **30**(6), 677–685 (1995)
80. S.M. Park, H. Yoo, 1.25-Gb/s regulated cascode CMOS transimpedance amplifier for gigabit ethernet applications. IEEE J. Solid-State Circuits **39**(1), 112–121 (2004)
81. C. Chan, O.T. Chen, Inductor-less 10 Gb/s CMOS transimpedance amplifier using source-follower regulated cascode and double three-order active feedback, in *Proceedings—IEEE International Symposium on Circuits and Systems*, 2006, pp. 5487–5490
82. H. Chen, C. Chen, W. Yang, J. Chiang, Inductorless CMOS receiver front-end circuits for 10-Gb/s optical communications. Tamkang J. Sci. Eng. **12**(4), 449–458 (2009)
83. M. Atef, H. Zimmermann, 10 Gbit/s 2 mW inductorless transimpedance amplifier, in *ISCAS*, 2012
84. E. Sackinger, *Broadband Circuits for Optical Fiber Communication* (Wiley, New Jersey, 2005)
85. M. Atef, R. Swoboda, H. Zimmermann, A front-end optical receiver for multi-level data transmission, in *Informationstagung Mikroelektronik* (*ME2008*), 2008, pp. 246–249
86. M. Atef, R. Swoboda, H. Zimmermann, An automatic gain control front-end optical receiver for multi-level data transmission, in *26th Norchip Conference*, Tallinn, Estonia, Nov 2008, pp. 57–60
87. M. Atef, R. Swoboda, H. Zimmermann,Optical receiver front-end for multilevel signaling. Electron. Lett. **45**, 12–122 (2009)
88. M. Atef, R. Swoboda, H. Zimmermann, Giga-bit optical receiver for plastic optical fibre. J. Opt. Commun. **283**(3), 391–395 (2010)
89. B. Offenbeck, W. Tschekalinskij, N. Weber, Eye safety in POF-systems—current view and future approaches, in *The 7th International Conference on Plastic Optical Fibers*, Santa Clara, USA, August 2008
90. M. Atef, R. Swoboda, H. Zimmermann, Optical receiver with large-area photodiode for multilevel modulation. J. Opt. Quantum Electron. **41**(2), 131–135 (2009)
91. T. Ridder, P. Ossieur, X. Yin, B. Baekelandt, C. Melange, J. Bauwelinck, X. Qiu, J. Vandewege, BiCMOS variable gain transimpedance amplifier for automotive applications. Electron. Lett. **44**(4), 287–288 (2008)
92. D. Cardenas, A. Nespola, S. Camatel, S. Abrate, R. Gaudino, The rebirth of large-core plastic optical fibers: some recent results from the EU project POF-ALL, in *Optical Fiber Communication and the National Fiber Optic Engineers Conference* (*OFC/NFOEC 2008*), March 2008, p. OWB6
93. M. Atef, W. Gaberl, R. Swoboda, H. Zimmermann, Multilevel signaling optical receiver for high-speed transmission over large-core step-index plastic optical fibre, in 17th *Austrian Workshop on Microelectronics* (*Austrochip 2009*), Graz, Austria, Oct 2009, pp. 1–4
94. M. Atef, W. Gaberl, R. Swoboda, H. Zimmermann, 4-PAM monolithic optical receiver, in *Proceedings of the Workshop Optics in Computing 2009*, Vienna, Sept 2009, pp. 48–52

95. O. Jeon, R.M. Fox, B.A. Myers, Analog AGC circuitry for a CMOS WLAN receiver. IEEE J. of Solid-State Circuits **41**(10), 2291–2300 (2006)
96. M. Atef, W. Gaberl, R. Swoboda, H. Zimmermann, An integrated optical receiver for multilevel data communication over plastic optical fiber, in *NORCHIP*, Trondheim, Norway, Nov 2009, pp. 1–4
97. M. Atef, R. Swoboda, H. Zimmermann, Optical receiver for multicarrier modulation in short-reach communication. Electron. Lett. **46**(3), 225–226 (2010)
98. J.M. Castillo, A.D. Sanchez, M.L. Aranda, Differential transimpedance amplifiers for communications systems based on common-gate topology, in *IEEE International Symposium on Circuits and Systems, ISCAS*, vol. 2, May 2002, pp. II-97–II-100
99. A. Kopa, A.B. Apsel, Common-emitter feedback transimpedance amplifier for analog optical receivers, in *IEEE International Symposium on Circuits and Systems, ISCAS*, May 2006, pp. 5482–5486
100. D.B. Nguyen, Q.H. Duong, T.A. Phan, H.H. Nguyen, X. Yu, S.G. Lee, A high linearity, low noise, and high gain transimpedance amplifier (TIA), in *8th International Conference on Solid-State and Integrated Circuit Technology, ICSICT'06*, 2006, pp. 1553–1555
101. J.P. Alegre, S. Celma, B. Calvo, N. Fiebig, S. Halder, SiGe analog AGC circuit for an 802.11a WLAN direct conversion receiver. IEEE Trans. Circuits Syst. II **56**(2), 93–96 (2009)
102. M. Guinea, T. Jouhti, M. Saarinen, P. Sipila, A. Isomaki, P. Uusimaa, O. Okhotnikov, M. Pessa, 622 Mbit/s data transmission using 650 nm resonant-cavity light-emitting diodes and plastic optical fiber, in *POF'2000*, Boston, Sept 2000, pp. 49–53
103. M.M. Dumitrescu, M.J. Saarinen, M.D. Guina, M.V. Pessa, High-speed resonant cavity light-emitting diodes at 650 nm. IEEE J. Sel. Top. Quantum Electron. **8**(2), 219–229 (2002)
104. S.C.J. Lee, F. Breyer, S. Randel, O. Ziemann, H.P.A. van den Boom, A.M.J. Koonen, Low-cost and robust 1 Gbit/s plastic optical fiber link based on light-emitting diode technology, in *Optical Fiber Communication and the National Fiber Optic Engineers Conference (OFC/NFOEC 2008)*, March 2008, p. OWB3
105. R. Krugloy, S. Loquai, O. Ziemann, J. Vinogradov, C. Bunge, Channel capacity of step-index polymer optical fibers: experiments and simulation with realistic parameters, in *Optical Fiber Communication and the National Fiber Optic Engineers Conference (OFC/NFOEC)*, 2010, p. JWA12
106. C. Gimeno, C. Aldea, S. Celma, F. Aznar, A cost-effective 1.25-Gb/s CMOS receiver for 50-m large-core SI-POF links. IEEE Photon. Technol. Lett. (99) 2012
107. J. Sundermeyer, J. Tan, C. Zerna, Integrated analogue adaptive equalizer for gigabit transmission over standard step index plastic optical fibre (SI-POF), Oct 2009, pp. 195–196

About the Authors

Mohamed Atef received the B.Sc. and M.Sc. degrees in Electrical Engineering, Electronics and Communications from Assiut University, Egypt, in 2000 and 2005 respectively. From 2006 to 2007 he was a researcher in Czech Technical University in Prague, Department of Microelectronics, working on the improvement of optical properties of quantum dots. He received the Ph.D. at Vienna University of Technology, Institute of Electrodynamics, Microwave and Circuit Engineering (EMCE), in 2010 and then has been doing post-doctoral research from 2011 to 2012 at EMCE of TU Vienna. Since 2010 he is an assistant professor at Assiut University, Egypt. His current research interests are in the area of analog integrated circuits, optoelectronic integrated circuits and short-reach communication over plastic optical fiber. He is author and co-author of around 30 scientific publications. Since 2012 he is senior member IEEE.

Horst Zimmermann received the Dr.-Ing. degree in the Fraunhofer Institute for Integrated Circuits (IIS-B), Erlangen, Germany in 1991. Then, Dr. Zimmermann was an Alexander-von-Humboldt Research-Fellow at Duke University, Durham, N.C., where he worked on diffusion in Si, GaAs, and InP. In 1993, he joined the Chair for Semiconductor Electronics at Kiel University, where he lectured optoelectronics and worked on optoelectronic integration. Since 2000 he is full professor for Electronic Circuit Engineering at Vienna University of Technology, Austria. His main interests are in design and characterization of analog deep-sub-micron and nanometer CMOS circuits as well as optoelectronic integrated CMOS and BiCMOS circuits. He is author of the two Springer books 'Integrated Silicon Optoelectronics' and 'Silicon Optoelectronic Integrated Circuits' as well as co-author of 'Highly Sensitive Optical Receivers'. Furthermore, he is author and co-author of more than 350 scientific publications. Since 2012 he is senior member IEEE.

M. Atef and H. Zimmermann, *Optical Communication over Plastic Optical Fibers*,
Springer Series in Optical Sciences 172, DOI: 10.1007/978-3-642-30388-3,

Index

4-level, 14, 19
4-PAM, vi, 7, 10, 11, 14, 19, 22, 44, 45, 50, 54, 55, 57–60, 90–96, 98, 100, 101, 105, 107, 110, 129
8-PAM, vi, 22, 45, 57, 59, 60, 99, 111, 129

A
Acceptance, 46
AC-coupling, 19
Active, 23, 24, 27–29, 51, 53, 54, 57, 68, 108, 112, 119
Adaptive, 2, 3, 24, 32–34, 36, 37, 39, 52–55, 57–60, 117, 118
Adaptive equalizer, 34
Agilent, 44
Algorithm, xix, 32–34, 37–.39, 55
AMGDM, 52, 55
Amplifier, v, vi, 2, 3, 24, 26, 27, 34, 55, 56, 58, 69–71, 73–76, 83–85, 88, 93, 97, 98, 100, 102–104, 112, 118, 119
Analyzer, 44, 54, 89–91, 121
Angle, 46, 47, 52
Anti-reflection, 67, 68, 103
Anti-reflection coating (ARC), 62, 63
Aperture, 41, 43, 46
Arbitrarywaveform generator (ARB), 16, 53, 91, 92, 100, 105
Area, xix, 1, 6, 7, 10, 23, 26, 28, 47, 51, 53, 56, 57, 60, 67, 68, 83, 95, 98, 100, 102–104, 108, 112, 117, 121, 123
Attenuation, v, 5, 22, 23, 42, 43, 45, 54, 57, 59, 60, 100, 126
Automatic, v, vi, 2, 67, 83, 84, 88, 89
Automatic gain control (AGC), xvii, 2, 57, 58, 60, 83–85, 88, 90, 96–99, 102–104, 107–109, 119
Automatic gain control (AGC) amplifier, 84
Automatic gain control (AGC)-TIA, 83, 119
Automatic power control (APC), 47
Automotive, 41, 42, 52
Average, 11, 12, 36, 37, 91, 110, 125
AWG, 19, 50, 53, 57

B
Background-light (BGL), 102, 103, 108, 109
Bandwidth, v–vii, 1–3, 5–10, 21, 22, 24, 29, 30, 42–48, 50–54, 56, 57, 60, 64, 67, 68, 69, 71, 73, 74, 78–81, 83, 84, 87, 90, 94, 95, 97, 99, 101, 105, 108, 110, 111, 114, 119, 120, 122, 124, 126, 129
Baseline, 19
Baud rate, 10, 50, 100
Bend, 3, 45, 52, 114
Bending, v, vi, 2, 41, 46, 117, 126
BiCMOS, v, vi, xvii, 23, 56, 67, 68, 78, 87, 98, 108, 120, 121, 123, 128, 130
Binary, vi, 2, 3, 5–7, 10, 11, 14, 19–22, 36, 44, 45, 50, 54, 57, 58, 88, 90, 92, 94, 98, 100, 105, 107, 110, 111, 115, 117, 122, 129
Bipolar, 67, 68, 75, 77, 103, 120
Bit error, 32, 57, 90, 124
Bit error rate (BER), vi, xvii, 3, 8, 9, 13, 23, 39, 40, 52–60, 85, 91, 94–98, 100, 107, 110, 111, 113, 114, 117, 118, 122–125, 130
BJT, 74, 77

M. Atef and H. Zimmermann, *Optical Communication over Plastic Optical Fibers*, Springer Series in Optical Sciences 172, DOI: 10.1007/978-3-642-30388-3,

B (*cont.*)
BPSK, 9
Bragg, 47

C
Cable, 1, 5, 120
Capacitance, v, 3, 30, 47, 67, 68, 70, 73, 77
Capacitor, 25, 34, 103, 118
Capacity, 5, 7, 8, 44, 46, 55
Carrier diffusion, 64, 65
Carrier drift, 64
CG-TIA, 80
Channel, v, xv, 5–11, 16, 17, 23, 24, 31–36, 38–40, 52, 53, 54, 110, 129
Chromatic, 48
Circuit, v, vi, ix, 3, 6, 14, 19, 23, 27, 28, 30–32, 36, 53, 54, 56, 58, 69–71, 74–76, 78, 79, 81, 83, 84, 87, 88, 96–98, 103, 104, 108, 118, 125
CMOS, xvii, 22, 23, 77, 78, 87, 103, 118, 120, 128
Combiner, 14, 31, 32
Common-base, xvii, 78–80
Common-emitter, 74, 75, 78, 103
Common-source, xvii, 27, 76, 81
Communications, v, vi, 1, 5, 8, 9, 20, 22, 43, 48, 90, 91, 103, 117, 129
Components, 2, 19, 20, 23, 24, 33, 34, 42, 54, 74, 118
Control, v, vi, 2, 26, 30, 42, 47, 73, 83, 84, 88, 99, 102–104, 118, 120, 121, 124
Coupling, vi, 19, 41, 46, 47, 53, 96, 97, 100, 101, 122–124, 126
Critical wavelength, 62
Current, 15, 27, 28, 30, 33, 36, 47, 49, 50, 69, 70, 74–77, 87, 98, 100, 103, 104, 119

D
D Flip-Flop, 14
Data rate (DR), v–vii, 1, 2, 5–8, 11, 20, 32, 36, 42–45, 48, 53–60, 65, 77, 90, 91, 94–96, 98–100, 105, 107, 110, 111, 117, 118, 122, 125, 126, 128–130
DC, 19, 29, 53, 56
DC-balance, 20
Decision feedback equalizer (DFE), 23, 36, 39, 40, 57, 58, 117, 118
Decode, vi, 2, 6, 19, 36, 37, 88, 98, 100
Delay, 5, 28–30, 32, 34, 54, 73, 83
Demodulation, 11, 52
Demultiplexing, 5, 53, 54
Diameter, 47
Differential, 9, 19, 22, 29, 30, 56, 83, 87, 103, 118, 119
Differential phase shift keying (DPSK), 9
Digital, 1, 2, 5, 52, 53, 57, 93–96, 100, 105, 111, 118
Digital-to-analog converter (DAC), 15, 52
Direct-Detection (DD), 5, 20
Discrete, v, 2, 3, 23, 30, 31, 38, 41, 43–45, 47, 56, 58, 87, 95, 100, 112, 113, 118, 124, 128
Discrete multi-tone (DMT), 2, 52, 53–55, 59, 60, 113
Dispersion, vi, 20, 22, 38, 42, 43, 46, 48, 50, 52, 53, 112, 126, 129
Distance, 10, 11, 20, 50, 53, 54, 60
Doping, 67, 68
Drift time, xvi, 66
Drift velocity, 66
DSP, 28
Dummy, 103
Dynamic range, 2, 56, 60, 82, 85, 88, 98, 103, 104

E
Electromagnetic interference (EMI), 1, 41, 42
Epitaxial layer, 67, 68
Equalization, vi, 2, 3, 22–24, 32, 34, 36, 37, 39, 52, 54, 55, 57, 58, 60, 97, 112, 114, 117
Equalized, 31, 39, 57, 118, 122, 124–126
Equalizer, v–vii, 2, 3, 23, 24, 26, 30–40, 52, 56–60, 112, 117–124, 128, 130
Error, v, vi, 3, 6, 7, 9, 11, 12, 33–37, 39, 53–60, 90, 94–97, 100, 101, 110, 111, 112, 114, 117, 118, 122, 129
Ethernet, 19, 22, 42, 56
Extinction ratio, 49, 50, 91, 92, 94, 96, 98, 102, 105, 107
Eye, 7, 14, 23, 36, 41, 50, 51, 54, 90, 91, 93, 94, 97, 98, 100, 101, 103, 105, 110, 112, 123, 126
Eye diagram, 14, 50, 90, 91, 92, 94, 97, 98, 100, 101, 103, 105, 110, 111, 112, 126
Eye safety, 95

F
Feedback, 23, 28, 30, 35, 36, 37, 39, 47, 71, 73, 74, 78, 80, 81–83, 97, 103, 104, 117, 118
Feed-forward equalizer (FFE), 57, 58, 112
FET, 76, 77

Fiber, v, vi, 1–3, 5, 11, 20, 22, 41–43, 46, 47, 50–53, 56, 57, 60, 92, 100, 105, 114, 117, 121
Fiber to the home (FTTH), 1
Filter, 24–30, 33–36, 52–54, 56, 58, 60, 98, 113, 118, 120, 124
Finite impulse response (FIR), xvii, 28, 30, 36, 60
Forward error correction (FEC), 57–60, 117
FPGA, 52, 57, 117
Frequency, vi, 3, 7, 9, 10, 19, 23, 24, 26, 27, 29, 34, 36, 39, 57, 71, 73, 76–78, 88, 107, 110, 111, 117, 120, 121, 124, 129
Frequency response, 34, 39, 43, 44, 52, 78, 89, 105, 110, 114, 117, 121–124, 129
Frequency-shift-keying (FSK), 9
Front-end, 77, 87, 88, 113

G
Gain, 73
Gaussian, 7, 38, 110
Gigabit, v, 14, 16, 22, 30, 55, 56, 97, 101, 120
Gilbert
GOF, 2, 41, 105, 110, 121
Green, 43, 60

H
Hamamatsu, 53, 54, 56–58, 60, 94, 95, 99, 112, 118
Heterojunction, 75
High pass, 24, 26, 27, 52, 120, 124

I
IC, 87
Implementation, vi, 7, 10, 27–29, 40, 42, 117, 118, 129
Inductor, 26, 27, 29
Infinite Impulse Response (IIR), xviii, 28
Information, 5, 6, 48, 85
Infrared (IR), 20, 48
Input capacitance, 70
Integrated equalizer, v–vii, 56, 58, 118, 122, 126, 128, 130
Integrated optical receiver, v–vii, 2, 3, 10, 54, 56, 65, 96, 97, 100, 101, 103, 104, 114, 117, 118, 121, 122, 124, 126, 128–130
Integrated photodiode, 61, 87, 88, 96, 99, 125
Intensity Modulation (IM), 5, 20
Intermodulation, xviii, 101, 103
Internet protocol television (IPTV), 1
Intrinsic layer, 65–67
Inverter with an active common-drain feedback (ICDF-TIA), 81, 82
ISI, 10, 23, 32, 33, 36, 38–40, 94, 107, 110

J
Jitter, 126

K
Kirk effect, 67

L
LAN, 5
Laser, xvi, xviii, 5, 10, 15, 16, 19, 22, 44, 47–54, 56, 57, 89, 91, 94–96, 98, 100–102, 105, 107, 117, 126, 94, 100
Laser diodes (LD), 44, 46–48, 50, 53–55, 58, 59, 96, 97, 101, 121, 122, 124, 125
Levels, v, 2, 6, 7, 10, 11, 14, 16, 22, 23, 40, 88, 90, 93, 100, 104, 105, 111, 126
Light emitting diode (LED), 46–48, 52, 56–60, 99, 113, 118
Light source, 48, 52, 56, 95, 111, 113
Limiting, v, vi, 2, 3, 28, 47, 83, 85, 88, 93, 98, 100, 110, 119, 120
Limiting amplifier (LA), 83, 84
Linear, v, 2, 23, 33, 36, 39, 40, 53–55, 58–60, 83–85, 88, 97, 98, 102–104, 109, 119, 129
Linearity, 2, 6, 54, 60, 93, 94, 98, 99, 102, 103, 105, 107–109, 111, 129
Link, 1, 20, 22, 42, 43, 48, 57, 60, 110
LMS, xviii, 32, 33, 39
Load, 26, 27, 29, 30, 75, 76, 104
Loss, v, vi, 2, 6, 7, 34, 41, 45–47, 51, 53, 60, 96, 97, 100, 101, 122, 126, 129

M
MATLAB, 39, 53, 91, 95, 97, 101, 105, 110, 111
Maximum likelihood sequence estimator (MLSE), xviii, 23, 36–40
MCML, 15
Meansquared error (MSE), 32, 33, 35, 36
M-levels, 6
Mode, 22, 34, 36, 42, 43, 46, 49, 50, 52, 53, 55, 100, 103, 105, 119, 126
Model, 52, 57, 58, 70, 74
Modulation, v–vii, 2, 3, 5–9, 21, 22, 32, 44, 47–50, 52–57, 60, 95, 98, 101, 109, 113, 117, 129, 130

M (*cont.*)
MOS, 26, 68, 120
MOSFET, xviii, 27, 30, 76, 77, 79, 83, 103
Multilevel, v, vi, 2, 5–7, 10, 15, 16, 19–22, 50, 85, 88, 90, 93, 95–100, 104, 114, 129
Multilevel pulse amplitude modulation (M-PAM), vi, 2, 9, 11, 16, 20, 22, 54, 57, 129, 130
Multiplexing, 5, 52, 54

N
Numerical aperture (NA), 43, 44, 46, 52, 53, 55, 95, 96, 100, 101, 111, 122, 124
Network, xix, 1, 21, 44, 89, 102, 121
Network analyzer, 105, 122, 124
Noise, v, vi, 3, 5–7, 8, 10–12, 22, 23, 30, 32–34, 36, 38, 39, 45, 47, 69–71, 74–80, 84, 87, 91, 94, 98, 100, 103, 105, 112, 119
Noise current, 69, 103
Nonlinear, 50
Nonlinearity, 10, 50, 60, 107
NRZ, 19, 53–55, 57, 58, 117, 118
Numerical, 41, 43, 46

O
Octave, 11, 44, 45, 129
OFDM, 101, 102, 104, 108
Offline, 53, 54, 57, 112
Optical communication, vi, 51, 39, 48, 50, 129
Optical modulation amplitude (OMA), 57
Optical power, xv, 3, 44, 45, 47–55, 57, 60–65, 85, 87, 91–98, 100, 101, 104, 105, 107, 110, 114, 119, 122–126
Optical receiver, v, vi, 2, 10, 19, 47, 54–56, 58–60, 67, 69, 70, 87–92, 95, 96, 98–100, 103–105, 107, 108, 110–114, 117, 118, 120, 122, 123, 124, 126, 129
Optoelectronics integrated circuits (OEIC), 87
Oscilloscope, 52, 53, 57, 93–97, 100, 101, 105, 110, 111
Overload, 82, 88, 119
Overshoot, 49, 73

P
Passive, 23, 24, 26, 28, 56, 117, 120
PC, 53
PCB, v, 3, 53
PCI, 50, 53, 100, 105
Peaking, 3, 23, 29, 52, 56, 58, 73, 95, 113, 117
Penalty, 6, 7, 10, 50, 94, 107, 110, 111, 129
Performance, v, 2, 3, 7, 10, 24, 28, 32, 33, 40, 42, 50, 54, 56, 57, 67, 70, 77, 87, 93, 99, 103, 117, 118, 129, 130
Phase, xvii, xviii, 7, 9
Phase shift keying (PSK), 9
Photocurrent, 61, 64, 65, 69, 88, 103, 109, 125
Photodiode (PD), v, vi, xv, xvi, xviii, 3, 47, 50, 52–54, 56, 58–62, 64–71, 74, 77, 79, 80, 87, 88, 94–96, 98, 100, 101, 103, 110–113, 117, 118, 122–125, 128, 130
Pigtail, 100
PIN, vi, xviii, 53–58, 66–68, 87, 88, 94, 95, 97, 100, 103, 112, 117, 118, 123, 126
Plastic optical fiber (POF), v–vii, ix, 2, 3, 22, 24, 39, 41–48, 50–60, 87, 95–97, 100, 101, 103, 110–112, 114, 115, 117–120, 122–126, 128–130
PMOS, 81, 104
Polymer, 41, 43, 65
Poly-methyl-methacrylate (PMMA), v, vi, vii, 3, 9, 22, 41–47, 52–55, 57–60, 95, 100, 101, 110, 111, 129
Poly-methyl-methacrylate plastic optical fiber (PMMA-POF), vi, 2, 42, 100, 129
Poly-methyl-methacrylate simple index plastic optical fiber (PMMA SI-POF), v–vii, 22, 45, 95, 96, 101, 111, 112, 114, 122
Post-equalizer, 23
Post-layout, 98, 105
Power, 2, 3, 6–10, 14, 22, 23, 28, 29, 32–34, 41, 46, 58, 87, 90, 91, 94–98, 100, 101, 103, 105, 107, 110–114, 117–119, 121, 122, 124, 126, 129
Power budget, 3, 51, 52, 57, 114, 117
Power consumption, vi, 54, 60, 79–82, 96, 128, 130
Power margin, 60, 122, 123
Power penalty, 7, 44, 50, 110, 111
PRBS, 53, 54, 90–92, 94, 96, 98, 100, 107, 110, 118, 122, 124–126
Preamplifier, 69, 70
Pre-distortion, 95
Pre-equalizer, 23, 52
Pre-equalizing, 3, 52, 56, 95, 113
Pulse amplitude modulation (PAM), v, vi, vii, 2, 9, 10, 22, 44, 50, 54, 56, 58, 60, 61, 91, 93, 94, 96, 98, 100, 101, 105, 107, 110, 112, 129, 130

Q
Quadrature amplitude modulation (QAM), 2, 9, 16

Quality factor, 12, 13, 83
Quantum efficiency, 63, 64, 68

R
Real time, vi, 1, 40, 53, 54, 60, 95, 96, 101, 111, 129
Receiver, 2, 3, 5–8, 10, 19, 23, 34, 35, 37, 45, 47, 53–60, 67, 69, 87, 93–98, 100–102, 104, 105, 107, 109, 111, 112, 118, 122, 125, 129, 130
Recursive least squares (RLS), xviii, 32, 33, 39
Regulated-Cascode TIA (RGC-TIA), 79–82
Relative permittivity, 65
Resistor, 26, 29, 30, 69, 70, 74–46, 103, 104, 120
Resonant cavity light emitting diode (RC-LED), 46–48, 52–55, 57, 58, 111, 112, 117
Response, vi, 2, 3, 7, 10, 23, 24, 30, 35, 38, 39, 117, 121
Responsivity, 63, 65, 69, 71, 103, 118, 125
Roll-off, vi, 6, 7, 11, 44, 45, 129
RS, 9, 57, 59, 60

S
Saturation, 66, 85
Sensitivity, v, 10, 41, 46, 47, 57, 59–61, 63, 65, 69, 70, 71, 78, 81, 82, 85, 90, 91, 94, 97–99
Settling time, 73
Set top boxes (STB), xviii, 42
Signal, v–vii, xv, xvi, xviii, 2, 5–8, 10, 11, 12, 14, 16, 19, 20, 22, 23–27, 30, 32–41, 44, 48, 50, 52–54, 56, 57, 60, 82, 83, 85, 87, 88, 90–105, 107–112, 114, 118–122, 124, 126, 129
Signaling, vi, vii, 5, 7, 10, 11, 20, 22, 44, 45, 56, 57, 90, 105, 110, 111, 112, 115, 129
Signal to noise ratio (SNR), xviii, 5, 7, 10, 12
Silicon, 23, 28, 56, 69, 118
Simple index plastic optical fiber (SI-POF), v–vii, 2, 3, 9, 22, 39, 42, 43–46, 52–60, 95, 100, 101, 111, 113, 114, 117, 118, 122, 124–126, 128–130
Simulated, 7, 107, 108
Simulation, ix, 97, 98, 103, 105
Small-signal bandwidth, 71
Small-signal transfer function, 70
Source, 3, 14, 16, 19, 24, 26, 29, 42, 44, 46, 48, 50, 67, 74, 77–79, 81, 89, 90, 96, 98, 100, 101, 105, 107, 118, 126
Spectral, xviii, 8, 32, 48
Spectrum, 10
Speed, v, vi, 3, 20, 23, 39, 47, 53–55, 57, 58
Step response, 73
Summing, 29, 30
Supply, 56, 58, 69, 83, 87, 88, 95, 98, 103, 104, 119, 120, 123
Symbol, vi, 5, 6, 11, 12, 16, 17, 22, 33, 34, 37, 38, 88, 94, 100, 102, 105, 110, 111, 118
Symbol error rate (SER), 11, 12, 91, 94, 100, 107, 110
Symbol rate, 5, 16, 17, 88, 105, 110, 111
Systems, vi, 2, 3, 5, 8, 9, 19, 20, 42, 50, 52, 54, 57, 60, 95, 102, 114

T
Tap, 28, 30–32, 34, 39, 58, 112, 118
Temperature, 42, 43, 47, 48
Threshold, 11, 47, 57, 60, 90, 98, 102
TIA, vi, xviii, 3, 30, 54–59, 70, 71, 74–83, 85, 87, 88, 97, 98, 102–104, 107, 112, 117–119, 130
Topology, 74, 103, 119
Transconductance, 76, 77, 79, 80
Transfer function, xv, 25–29, 33, 72, 73, 83, 85, 118, 124, 126
Transimpedance, v, 3, 30, 47, 52, 53, 60, 69, 72–74, 87, 88, 94, 97, 99, 103, 113, 119
Transimpedance amplifier, v, vi, 3, 47, 52, 53, 60, 69, 73, 74, 87, 94, 99, 113, 119
Transimpedance gain, 30, 70, 79–82
Transmission, v–vii, ix1–3, 5–7, 19, 20, 22, 35, 42, 43, 46, 52–60, 63, 90, 95, 97, 99, 100–102, 110–114, 117, 120, 122, 124, 126, 129, 130
Transmitted, v, 2, 5, 9, 11, 20, 31–33, 37–39, 43, 44, 52–54, 60, 90, 91, 95, 92, 100, 101, 105, 110, 111, 125, 126, 129
Transmitter, 5, 6, 9, 10, 15, 22, 23, 34, 50, 52, 54, 56, 57
Transversal, 27, 28
Trigger, 17

V
Variable gain amplifier (VGA), 56, 58, 81, 118
Vector signal analyzer (VSA), 54
Vector signal generator (VSG), 53
Vertical Cavity Surface Emitting Laser (VCSEL), 26, 46, 48, 53, 55, 58, 90, 92, 94, 95, 117, 123
Visible range, v, 2, 41
Viterbi, 37–39

V (*cont.*)
Voltage, v, 2, 6, 7, 14, 15, 26, 27, 29, 30, 67, 69, 74–76, 83, 88, 91, 96, 98, 100, 103–105, 111, 118

W
Waveforms, 5, 53
Wavelength, 5, 44, 46–48, 54, 60, 67–69, 90, 91, 94, 100, 107
Wavelength division multiplexing (WDM), 5
Weights, 31, 34, 39
Wireless, xix, 1, 102
Wireless local area network (WLAN), 1

Z
Zero, xix, 11, 29, 32–34, 49, 57, 103

MIX
Papier aus verantwortungsvollen Quellen
Paper from responsible sources
FSC® C105338

If you have any concerns about our products,
you can contact us on
ProductSafety@springernature.com

In case Publisher is established outside the EU,
the EU authorized representative is:
Springer Nature Customer Service Center GmbH
Europaplatz 3, 69115 Heidelberg, Germany

Printed by Libri Plureos GmbH
in Hamburg, Germany